The Bluebeam Guidebook

The Bluebeam Guidebook

Game-Changing Tips and Stories for
Architects, Engineers, and Contractors

Rachel Attebery

Jason Hascall

Library of Congress Cataloging-in-Publication Data:

Names: Attebery, Rachel, 1990- author. | Hascall, Jason A., author.
Title: The Bluebeam guidebook : game-changing tips and stories for
  architects, engineers, and contractors / Rachel Attebery, Jason Hascall.
Description: Hoboken, New Jersey : Wiley, 2018. | Includes index. |
  Identifiers: LCCN 2017056760 (print) | LCCN 2017056891 (ebook) | ISBN
  9781119393962 (pdf) | ISBN 9781119393955 (epub) | ISBN 9781119393948
  (paperback : acid-free paper)
Subjects: LCSH: Building information modeling–Computer programs. | PDF
  (Computer file format) | Bluebeam Revu. | BISAC: TECHNOLOGY & ENGINEERING
  / Construction / General.
Classification: LCC TH438.13 (ebook) | LCC TH438.13 .A88 2018 (print) | DDC
  720.285/53–dc23
LC record available at https://lccn.loc.gov/2017056760

Printed in the United States of America

SKY10073289_041824

To my wife, Dr. Jenny Johannes-Hascall, for her consistent support and persistence in demonstrating to me that we are each capable of more than we imagine.

—Jason

Always, First, and with Love, the time and effort and any success of this book is dedicated to my Savior, Jesus Christ.

Always, Second, and with Love, this book is dedicated to my husband Aaron who is made of the ingredients of support, understanding, and partnership.

Always, Third, and with Love, this book is dedicated to my family and close friends who have been excited for me and supported me and "that Bluebeam thing" even though they never really knew what it was.

And finally, this book is dedicated to every person with a compelling sense of "surely there must be a better way." You're right. I hope this book gives you some tools, and I hope to learn something new from you soon.

—Rachel

Contents

Foreword

Innovation is not an end in and of itself. Its value is derived from its application and mass adoption—otherwise, it has no context, no sustainability, and thus no meaning. Let's consider this statement—context is to exist in a place; sustainability is to persist over time; and meaning is to influence culture. To exist in a place, innovation must resolve its value within an industry sector(s). To stand the test of time, innovation must inspire long-term change and transform behavior. To influence culture, innovation must be at scale to shift a community from what is to what could be. Therefore, the value of innovation is rooted in its impact on the many and not the few or the one.

Logically, to promote adoption by many individuals, one must overcome natural barriers to change, which in a knowledge-centric economy are the new, the unknown, and the uncomfortable. The solution to breach these change barriers is mass education—education to promote a better understanding of technology and to empower individuals to adapt to change, not fear or oppose it. This is what Jason and Rachel provide—first-hand knowledge from the context of AEC project experience with the goal to educate and in so doing, propel the community forward.

Through practical, step-by-step instructions and case studies, this guide will allow individuals to capture the often-elusive value of technology—its application and adoption in the marketplace. The stories told within the case studies alone are knowledge gold as we consider the amount of time it would take to learn from successes and failures without standing on the shoulders of Jason and Rachel and the others who contributed their stories to this book.

Ultimately, this knowledge transfer will allow teams to move up the learning curve quicker and focus their efforts on value-added work, not the trivial or mundane. Rachel and Jason clearly understand what is driving the industry today—commoditization of technology and obsolescence of human capital—and have concluded that to transform the AEC industry, knowledge must be open sourced, and real value is the impact of networked individuals working together, not simply the results from individual performers.

This is a magnanimous position to take as they have effectively captured years of experience into a guidebook designed to support collaborators and competitors alike. Could this generosity be rooted in the position that they will grow from the experience of knowledge sharing and to improve, they must embrace competition and continuously redefine their comparative and competitive advantages? Maybe.

Regardless, by shifting up the learning curve, one of Rachel's goals for the industry is met—to transfer the knowledge of more experienced professionals to the next generation and potentially fill the ever-growing knowledge gap between generations. However, her ultimate goal goes beyond knowledge transfer to the intersection of human experience with artificial intelligence (AI)—to improve cognition of learning machines. This vision might fly in the face of Jason's concern that the human component of work might be rendered irrelevant with advanced AI, but peering into the future, one can also postulate that AI would simply capture another layer of knowhow, pushing the human experience higher up the value chain. Although seemingly aspirational, the next five years will bring much of this to reality as we project the evolution of IBM Watson and the metamorphosis of learning algorithms into cognitive algorithms that contextualize data, providing for direct application of information to the built world.

Ultimately, technology's impact on society is directly related to the speed at which man moves up the learning curve—human capital must grow in direct proportion to the speed of technology. Work today is simply different than in the past—it has not been eliminated—it has evolved. Arguably, the future holds much of the same with the added benefit, as Rachel would proffer, of a higher quality of life. Jason would hold the position that the introduction of new technologies could positively or negatively impact the world, which underlies his concern. However, these opposing forces are what also bring a level of excitement and motivation for Jason to be at the forefront of technology, so that he might demonstrate how to positively leverage technology for a better future.

Within these pages, Jason, an experienced structural engineer who sees himself as calming, cool, and bold, and Rachel, an experienced chemical engineer who describes herself as specific, deep, and imaginative, provide their perspectives and extensive experience with Bluebeam technology. This guidebook is thorough and creative. Their stories are what bring the technology to life and lift you to a level of understanding that will change your world, one written word at a time.

When I was the CEO of Bluebeam, Inc. I had one goal in mind—to change the world by helping people do what they do better. This rudimentary concept is based on the idea that change management is the most difficult challenge one must overcome in order to adopt technology. And, that to promote change, the goal is not to change processes, but to enable them with technology and education. In doing so, efficiencies would be gained in the transformation of processes through the acceptance and adoption of technology. This philosophy allowed Bluebeam to average a 50% compounded annual growth rate for 15 years and expand its footprint to well over one million users worldwide in 130 countries.

As a technologist, a member of this community, and a staunch advocate of collaboration, I would like to thank Jason and Rachel in earnest for this knowledge guide to a better future.

—Richard L. Lee
Bluebeam Founder and CEO
2002–2017

Acknowledgments

To the daring professionals and colleagues who were bold enough to share the stories that make up the case studies within this book, thank you. It is your contributions that make this endeavor possible and your willingness to share that supports the resounding theme of the text within.

To the many professionals at Bluebeam who created a software that changed the AEC community and simplified my life, thank you. You listen like no others; your sense of community is unsurpassed; and your creativity is something to be marveled at. It is your creation that gives this book purpose and meaning.

To Larry Naab, who introduced me to Bluebeam, which has not only influenced my career but also the direction of our entire company, thank you. I can't imagine a week without Revu.

To my co-author, Rachel Attebery, who jumped into this endeavor without a moment of hesitation, thank you. I am proud of what we accomplished.

To Brad Hardin, who said, "Go for it!" thank you. Your leadership has opened doors that wouldn't have been opened otherwise.

And finally, most importantly, to Jenny and Truman, thank you. Your patience, encouragement and support, not the least of which was the occasional candid reminder to "FINISH THE BOOK," carried me to the finish line. You both fill my life with joy and purpose every day, and I can say with 100 percent certainty that I would never have become an author without you.

—Jason

I'm so grateful to Black & Veatch for giving me the chance to try something new.

Jason Patterson was the first person to introduce me to Bluebeam, and he did so against many odds. Thanks for changing the course of my life, Jason.

Jason Hascall was the first person to explain Bluebeam to me. His passion, clarity of thought, and creativity continue to motivate me to be better.

Brad Hardin introduced us to his contacts at Wiley and encouraged us to write this book. Without his support and connections, this book wouldn't be here today.

The Black & Veatch Bluebeam UserX team, started by Brad Hardin, has become family. You all impress me every time we get together, and some of you are featured in this book. Thank you for your irrepressible sense of "it could be done better."

The contacts we've made through the Kansas City Bluebeam User Group have made us feel part of a Bluebeam community in Kansas City. Thank you for your contributions to this book, the ones that are obvious and the ones you will never know.

The company of Bluebeam itself seems to be made only of enthusiastic, passionate, hugging, name-remembering rockstars who have supported us no matter how dumb our questions are. Thanks for setting a new standard of software service and pushing the envelope on useful software.

—Rachel

Introduction

What Is This Book About?

This book is a Rosetta Stone for anyone in the architecture, engineering, and/or construction industries who wants to implement new technology but avoid making beginner's mistakes on a real project with real budget and real deadlines. Bluebeam Revu is a powerful PDF software made specifically for our industry, and it is the featured technology in this book. Its creators are people like us, who have lived the painful day-to-day slog of AEC workflows and did something about it.

As a company, Bluebeam focuses on innovatively streamlining workflows for these tasks:

- Design reviews
- Bids and estimates
- Requests for Information (RFIs) and submittals
- Field inspection
- Site management
- Punch and back check
- Closeout
- Facilities management

As of 2017, they achieved the following customer base:

- 100% of Top 50* Contractors
- 98% of Top 50* Design Firms
- 100% of Top 50* Design/Build Firms

It is unusual for professionals in our industry to share lessons learned with competing firms, but this book doesn't care about that. After all, if you've figured out something great, you're probably not the first one. Why not share and get new information in return? This book uses three approaches to illustrate best practices for making the most of your document-based workflows with Bluebeam:

1. Interviews real AEC professionals for success stories
2. Explains click-by-click how you can be successful too, with contextual lessons learned
3. Shares exclusive expert tips from power users

---

* Top 50 firms by revenue as reported by Engineering News-Record

Who Is This Book For?

This book is for anyone in the AEC industry, including design and field engineers, contractors, project managers, architects, CAD technicians, equipment/material vendors, construction workers, administrative assistants, document control personnel, and everyone else who wants to improve the way their project's drawings and documents are handled. Within the tasks listed above, Bluebeam improves specific processes traditionally done on paper, such as preserving the original vector quality of drawings and documents all the way through to their final delivery; signing documents digitally from anywhere in the world; creating real-time, global collaboration for markups; making drawings available digitally in the field; storing 360° site photos within a design drawing; estimating quantities and takeoffs; viewing the design model in a 3D PDF; and much, much more.

This is the perfect time to learn how to use these features to your advantage. A 2016 survey of AEC professionals around the world showed "a strong relationship between investment in advanced technologies and performance vs. competitors" (CIMdata, www.cimdata .com/en/resources/complimentary-reports-research/commentaries/item/7090-aec-technologies-and-transformation-survey-results-and-interpretation-commentary). Especially over the past four years, increased spend on IT correlates to higher and even leading performance.

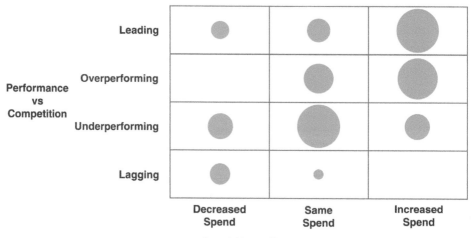

Company Performance vs. IT Spend

CIMdata, "AEC Technologies and Transformation: Survey Results and Interpretation (Commentary)," Oct. 26, 2016

By the way, this is an excellent survey summary to read to understand the technology labor pains our industry is in right now. The digital tide swept the world decades ago, and our industry has ridden the waves to some extent. Most of us have computers at least. But when it comes to really overhauling the way we work to embrace the latest tools on the market, AEC is traditionally . . . well, traditional. Change is difficult but necessary to survival. If you and your company can be open-minded enough to entertain thoughts of challenging "the way we've always done it," you have great odds to lead our industry and differentiate yourself from the competition. Our collective struggle to leverage technology is rebutted by owners who would still rather receive a cardboard box of paper drawings. Fine—you can still print paper deliverables for whomever wants them; but that doesn't mean you have to operate internally with hammers and chisels. Take the efficiency gains for yourself, rally your company around a better method, and, as always, serve the clients whatever they ask for. The customer is king, and they will be just as happy to accept their paper drawings in half the time and at twice the quality.

What Benefits Will I Gain by Reading This Book?

As you read this book, you will learn if Bluebeam Revu is a useful tool for you. If that proves to be the case, you'll hear the true stories of other Bluebeam users, learn the how-tos of Bluebeam's most powerful features, and profit from the recorded mistakes made by early adopters. You'll be able to champion Bluebeam Revu within your company, armed with the collective knowledge of power users across many firms. You will also gain inspiration from the nuggets of users who persisted in their quest for a better way; you are certainly not alone in this journey, which is bigger than one software tool. The classic story of a few lone evangelists with a message from the future is woven through this book, and we hope you'll feel camaraderie and support to keep going.

Why Did You Write This Book?

This is the only book of its kind on the market today. In the face of furtive, competitive secrecy, we've decided to bare it all for the sake of implementing new, innovative technology at historically entrenched AEC firms. The authors work at a century-old engineering, procurement, and construction company and were among the first to adopt, and moreover promote the use of, this new technology. In our saga of experimenting, talking with peers at other companies, navigating cultural resistance, and building a sustainable support structure, we've learned that sharing information only benefits its sharers. Without an exchanged flow of experiences, we as an industry will stagnate. Let's gain our competitive edges by doing amazing, trustworthy, genius work and not by hoarding technology successes. Publishing this information is a duty we have to ourselves, and to you, as cohorts in this great industry of architecture, engineering, and construction. We wish you well in all your endeavors and hope this book serves you in a useful way. On to chapter 1!

Chapter 1
Taking the Leap: Switching from Red to Blue

Blue isn't for everyone. To understand if Bluebeam might be the right choice for a group or company, it helps to understand how the increasingly popular PDF tool came into being. As evidenced by the tool's devotion to efficiency, it was the brainchild of a team of engineers seeking to make their own work faster and easier. In the late 1990s, NASA's Jet Propulsion Laboratory was tasked with designing and building the robotic arms for the Mars rovers *Spirit* and *Opportunity*. They only had two years to do it, and with a little fewer than 20 employees, they had to get creative. Although Adobe was running the PDF market at that time, their products didn't meet the specialized needs of the engineering team, specifically, creating accurate PDF files from computer-aided design software such as AutoCAD. Thus, the inception of Bluebeam's first product: Pushbutton PDF. The software continued to evolve and become more popular, and Brett Lindenfield, the director of the robotic arms project, called on his friend Richard Lee to turn what started as a humble innovation into a full-blown company. Pushbutton PDF spun off into its own company in 2002 as Bluebeam Software Inc. The product began to diffuse from the engineering industry into architecture, construction, and oil and gas, and was renamed Bluebeam Revu. Read the full story on NASA Spinoff: https://spinoff.nasa.gov/Spinoff2016/it_3.html. Fourteen years later, in 2016, Bluebeam Software Inc. achieved 1,000,000 users in over 100 countries and is used by 98 percent of the world's top 50 design-build firms by revenue.

Products and Feature Comparison

Even with over a million users requesting new features, Bluebeam has stayed true to its initial industry, focusing on solving problems for design reviews, bids and estimates, RFIs (request for information) and submittals, field inspection, site management, punch lists and back checks, project closeout, and facilities management. Bluebeam products are not designed for artistic use in the same way that Adobe Creative Cloud or Photoshop or Illustrator are. They are, however, highly competitive with Adobe's base products, such as Adobe Reader and Adobe Acrobat. Like Adobe, Bluebeam has a portfolio of products tailored to specific needs. As of the time this book was authored, here are Bluebeam's offerings:

- Bluebeam Revu: Bluebeam's flagship product, built for Windows PCs and tablets. Bluebeam Revu is a PDF workhorse. It comes in three flavors: Standard, CAD, and eXtreme, in order of increasing price.

- Bluebeam Revu Mac: Most of the functionality of Bluebeam Revu, but for Apple PCs
- Bluebeam Vu: Free PDF view-only tool
- Bluebeam Revu iPad: Mobile version of Revu
- Bluebeam Vu iPad: Free mobile version of Vu

As stated above, although Bluebeam is a fantastic tool, it is only fantastic when it's the right tool for the job. Remember the NASA engineers—they weren't out to create a PDF tool for everyone; they just wanted something to meet their specific needs. To understand which product is the right fit for a given use case, consider the comparison chart of high-level features for Bluebeam and Adobe products (Table 1-1). For a more detailed comparison, see www.bluebeam.com/us/products/revu/compare.asp, www.bluebeam.com/us/bluebeam-difference/bluebeam-vs-adobe.asp, and www.bpsboise.com/wp-content/uploads/2013/12/Bluebeam_vs_Adobe.pdf.

License Pricing

After deciding which product offers the best feature set for a given group, the next consideration is cost. For current pricing, visit Bluebeam's and Adobe's websites. As with many products, the bigger the enterprise is, the better the volume discounts and included services will be—for both Bluebeam and Adobe. To get the best and most accurate pricing, call both software providers and talk through the specifics of what is needed.

Bluebeam has developed considerably in its enterprise licensing options, now offering the choice of perpetual licensing or open licensing for Revu eXtreme only. Typical perpetual licensing registers one seat of product per workstation, while open licensing is cloud-based and allows all users to share licenses on an as-needed basis. For example, if an organization owns 100 open licenses of Revu eXtreme, up to 100 people can be using Revu eXtreme at the same time, and when one user closes his or her instance of the software, a different user can now open it. With perpetual licensing, there can be only as many total users as there are purchased licenses. Therefore, open licensing is useful when this organization has over 100 users *total* but no more than 100 *simultaneous* users. Unlike perpetual licensing, which is a one-time purchase, open licensing is paid on a recurring annual basis (Figure 1-1). Open licensing may be a worthwhile option for a company operating out of several different time zones where the cost savings of reduced license count outweighs the cost of annual subscription. Companies should consider how many years they plan to use Revu eXtreme and then calculate the total cost of ownership for the projected number of required licenses to decide whether open or perpetual licensing is the right choice. Carrying out the example of an organization with 100 open licenses, the chart below shows that it will take five years with 850 total users for the cost of open licensing to become more expensive than perpetual licensing, given the example pricing in Table 1-2. This pricing is for illustrative purposes only; please see Bluebeam's pricing website for accurate costs: www.bluebeam.com/solutions/pricing/.

For more information about open licensing, see www.bluebeam.com/us/_media/pdfs/DM-OpenLicensing-CMYK-Nem-Mech.pdf.

Table 1-1: Bluebeam and Adobe Features

	Bluebeam Revu Standard	Bluebeam Revu CAD	Bluebeam Revu eXtreme	Bluebeam Revu Mac	Bluebeam Vu	Adobe Acrobat Standard DC	Adobe Acrobat Pro DC	Adobe Acrobat Reader DC
Redline PDFs	X	X	X	X	In Studio	X	X	
Create custom markups	X	X	X	X				
Track annotations	X	X	X	X	X	X	X	X
Online collaboration	X	X	X	X	X	X	X	X
Pin files	X	X	X		X			
SharePoint integration	X	X	X		X	X	X	X
ProjectWise integration	X	X	X		X			
Markup 3D PDFs	X	X	X	X	In Studio	X	X	
Create 3D PDFs		X	X				X	
Create PDFs from Microsoft Office	X	X	X			X	X	
Create PDFs from any Windows file	X	X	X			X	X	
Create PDFs from AutoCAD		X	X			X	X	
Create PDFs from Revit, SolidWorks		X	X					
OCR		X	X			X	X	
Form creation		X	X				X	
Redaction		X	X				X	
Scripting		X	X				X	
Measure PDFs	X	X	X	X	In Studio	X	X	
Flatten markups	X	X	X	X		X	X	
Markup layers	X	X	X	X		X	X	
Markup summary	X	X	X	X		X	X	
Import markups	X	X	X	X				
Calibrated markups	X	X	X					
Create hyperlinks	X	X	X	X		X	X	
Dynamic stamps	X	X	X			X	X	
Embed photos and video	X	X	X	X	In Studio			

(continued)

	Bluebeam Revu Standard	Bluebeam Revu CAD	Bluebeam Revu eXtreme	Bluebeam Revu Mac	Bluebeam Vu	Adobe Acrobat Standard DC	Adobe Acrobat Pro DC	Adobe Acrobat Reader DC
Translate markup language	X	X	X					
Custom calculations	X	X	X					
Custom statuses	X	X	X					
Markup legends	X	X	X					
Document tags	X	X	X					
Extract, delete, rotate, insert pages	X	X	X	X		X	X	
Embed file attachments	X	X	X	X		X	X	
Overlay pages	X	X	X	X			X	
Combine PDFs	X	X	X	X		X	X	
Reduce file size	X	X	X	X		X	X	
Split document	X	X	X			X	X	
Headers and footers	X	X	X	X		X	X	
Create bookmarks and page labels	X	X	X	X		X	X	
Text search	X	X	X	X	X	X	X	X
Visual search	X	X	X		X			
Custom toolbars	X	X	X		X	X	X	
Color processing	X	X	X			X	X	
Batch functions			X					
Password protection	X	X	X	X		X	X	
Create digital signature fields	X	X	X					
Sign digitally	X	X	X	X	X	X	X	X

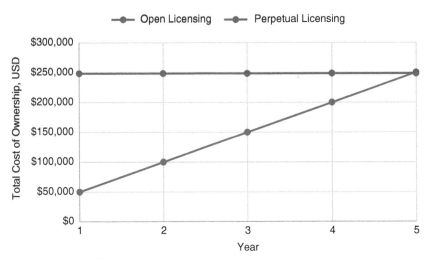

Figure 1-1: Cost of Open Licensing versus Perpetual Licensing

Table 1-2: Open versus Perpetual Licensing Example

Licensing Type	Number of Licenses	Price per Seat	Maintenance	Upgrade	Total Cost per Seat
Open licensing	100	$500	Included	Included	$500/yr
Perpetual licensing	850	$175	$30	$87	$292

Revu eXtreme is currently the only Bluebeam product with the option of open licensing. All Revu products, with the exception of Mac and iPad, offer volume discounts on a transaction-by-transaction basis. Pricing is different for Revu Standard, CAD, and eXtreme, depending on whether the purchase location is the United States and Canada or elsewhere in the world.

Unlike Bluebeam, Adobe's volume discount pricing is not clearly stated on its website. To attempt to compare costs as "apples to apples," compare the prices for individual licenses of Adobe Acrobat Standard DC and Pro DC on an annual prepaid basis, as shown on Adobe's website (https://acrobat.adobe.com/us/en/acrobat/pricing.html?promoid=KLXMR), to the prices of individual perpetual licenses of Bluebeam Revu Standard and Bluebeam Revu eXtreme in the United States, without including maintenance or upgrades.

Value Proposition

Because Bluebeam offers so many features in its products, it can be difficult to decide if the return on investment will be worth spending more money on a more advanced edition. After a base edition is selected based on the group's must-have features, next look at the "nice-to-have" features and figure out how much time and money would be saved by being able to utilize those features. For example, an organization isn't sure if it should pay the extra money for Revu eXtreme instead of the basic Revu Standard. Pick a "nice-to-have" feature, like OCR (optical character recognition). OCR allows scanned paper documents to be processed into searchable PDFs. Do a quick case study on the value of OCR by asking a professional how often he or she searches PDFs, how long it takes to find content in a nonsearchable PDF, and how long it takes to find content in a searchable PDF. One such case study showed a 54 percent reduction in time to find the desired content when using an OCRed document instead of the original scanned document. If an engineer is currently spending two hours per week searching scanned documents and works at a rate of $55 per hour, the cost savings equate to $59.40 per week, and the cost difference between Standard and eXtreme pays for itself in less than a month.

This OCR example is one data point in a bigger technology selection philosophy. It's shortsighted to make an enterprise purchasing decision based only on must-have features, because all too often, the must-have features are actually just a copy of how things are currently done. If the switch to a new product just mimics the capabilities of the old product, that's a miss. Technology is moving so rapidly that each product change should be accompanied by a new suite of features that makes a business better, faster, stronger. It's unfortunate that in many large organizations, the person making decisions about enterprise technology is not usually the user of that product. To get the best possible value out of a new product, the technology decision maker should spend time talking with the end users of the potential product and understand what they like and don't like about their current product. What do they like being able to do? What would make their jobs easier or let them do more in the same amount of time? Also think about the future of the company—what goals have executive leadership set for the next 10 years? What does company growth look like, in terms of employee count and global presence? What new markets might the company enter? What types of new professionals will the company hire? How does the company hope to do business in the future? What tools would be important to keep the company at the leading edge among its competitors? Because technology is embedded in every part of life, it has a powerful impact on a company's success, even its survival. Technology decision makers have more responsibility than ever before to understand their company's big picture objectives and make future-looking choices to help achieve those goals. The days of simply "keeping the servers up" are over. The most valuable information technology professionals will be those who can understand the *future* of business and make decisions to prepare their organization for that impending future.

Meet Larry Naab

Electrical Engineer, Black & Veatch

Change isn't always easy, and change at a 10,000-person engineering firm can feel downright daunting. The process for approving the change may take longer, and require more effort than actually implementing the change. It can wear down even the most resilient of professionals. But still some of those changes happen and some of them have big impacts that alter a company forever.

That was the case for Larry Naab, an electrical engineer at Black & Veatch. In early 2012, Larry was serving as the design manager for a large project. While attending a recurring

Project Management Lunch 'n' Learn, he was introduced to a product called Bluebeam Revu. The Black & Veatch project manager giving the presentation had come from another firm in the area and cited the remarkable benefits of using Revu. He claimed that it was so good that he purchased a copy to use at home on a personal level.

Larry watched as the PM showed the basic features of Bluebeam and shared how each feature had made a difference at his previous employer. He showed markups and markup tracking, measuring, searching, and an array of other capabilities. Larry liked the features; they were no doubt better than any electronic quality control software he had seen, but he was still a skeptic. Electronic quality control reviews had been tried, but none had been successful. At the end of the day, it was faster to work on paper and professionals would abandon the electronic way in favor of the hard copy from which they would then copy their marks back into the electronic system. Could Bluebeam be any different? The project manager said, "Yes."

With an upcoming 65 percent quality control review on his project, Larry saw an opportunity and set out to try out this Bluebeam for himself. If he could convince the project team to use Revu, he would not only save the reproduction costs of printing the hard copies, but also allow two to three more days of design and review time, something that Larry didn't take lightly.

In Larry's mind, a big reason why electronic quality control reviews weren't successful was because of the dismal screen real estate available to a reviewer. It had been nearly impossible to see the whole drawing at a time while still maintaining the ability to read the text. The easy solution in Larry's mind was to utilize a larger monitor and so he requested a 24-inch variety to test the capabilities of Revu himself. If Larry wanted the project team to use Revu, he personally had to want to use Revu.

The request for the larger monitor was quickly declined, noting that exceptions for one professional couldn't be made without causing hard feelings for others. Determined that Bluebeam could make a difference, Larry didn't stop there. He went to the local electronics store, purchased a 24-inch monitor for himself, and showed up to work the next day with a new piece of hardware. Little did Larry know what he was about to start.

With a free trial copy, Larry set off testing Revu. He changed his default PDF software to Bluebeam so any PDF would open in Revu without any additional effort, forcing himself to actively use the new software. To this day, Larry claims that switch was the most important step in his Bluebeam conversion.

Larry tried out numerous features, from markup and measuring to comparing documents and text recognition. He even accessorized with a stylus and touchpad, which he ended up completely disliking.

Though some capabilities lagged behind its competitors, in Larry's mind, Bluebeam came out on top overall. This was something he could use. This was something his project could use. This was something Black & Veatch could use.

For his immediate need, Larry believed he could shave his quality control cycle by 3 days if the project utilized Bluebeam. That in turn would extend the design time of the project at the time when it needed it most. The challenge was getting the 20-person project team to agree on going electronic. Larry refused to force or mandate his team to use Bluebeam, citing numerous failures in other software rollouts due to a lack of professional buy-in. Instead, Larry needed them to want to use Bluebeam. He needed an incentive and he knew of just the thing, larger monitors. Larry struck a deal with the project team; anyone agreeing to use Bluebeam Revu exclusively for the next QC cycle would receive a large, 24-inch monitor paid for by the company, essentially eliminating screen real estate issues.

"Do we get to keep the monitor?" "Can I use the monitor on other work?" "What if we don't like Bluebeam?" Larry heard all the questions. He was so confident in Bluebeam's success that he agreed, use Revu on this one project and the monitor is yours to use on all future work. No further commitment on using Bluebeam, no further strings attached. The decision was nearly unanimous and Larry put his plan into action by hosting an hour-and-a-half introduction to Revu.

At 65 percent, the Bluebeam debut, Larry hit a few snags. A few professionals demanded paper, so he printed a handful of hard-copy sets. A couple professionals groaned that Bluebeam was a memory hog and made their computers run slow, so Larry printed a few more. Conflicts with access rights were causing delays, so Larry created an individual electronic file for each professional. But after that, the Bluebeam experiment was running smoothly.

The team used Bluebeam for the two remaining review cycles, and following project completion, Project Management held their usual continuous improvement meetings, or the "Lessons Learned Rundown." The response was unexpected and overwhelming; the team loved Bluebeam, citing the benefits of the measurement tools, the searching tools, and the callout tool. Professionals said, "I have to have this now, all the time." "I want to use this on every project." "I don't ever want to go back to the old way."

Larry said, "I certainly didn't expect it to go THAT well!"

At this point, word was out. It seemed like everyone in the Business Unit wanted to get Bluebeam and everyone wanted it now. Project managers were asking Larry how to roll it out on their projects. Large groups were asking Larry for demo sessions. Engineers and technicians were asking how to get larger monitors. And then it happened; the floodgates opened, leadership said, "Okay."

Larry worked with Steve Mitts, the chief engineer at the time, to determine which version of Bluebeam Revu was appropriate and how many licenses the division should buy. They also hashed out an official Electronic Quality Control process to create some uniformity between projects. Together they decided on 50 additional licenses of Bluebeam Revu eXtreme.

Larry noted that the tipping feature on the eXtreme decision was optical character recognition, or OCR, the ability to convert nonsearchable text to searchable text. Though technology allows the ability to create PDF files with searchable text, many of Larry's clients were not doing that on their own. The ability to search a document was too valuable to let it go.

The growth continued organically, one project at a time. After 25 additional licenses, 50 more, and then, just one year later, Revu became the default PDF software for the entire Business Unit. Everyone got it on their computers as the standard.

As the growth continued, so did the learning. Larry's original setup was exactly like paper, not utilizing many of the benefits afforded by the new electronic process. Redlines were statused on the page with supplemental redlines, the markup tracking capabilities weren't utilized at all, and filters weren't even discovered yet. But slowly, that all changed.

With each project came new professionals exploring how to best utilize the new tool. Teams discovered ways to allow multiple reviewers to review simultaneously. They learned how to utilize the markup list as the "to-do" list. They discovered that filters were one of the best features they didn't even know they needed.

According to Larry, the way his business uses Bluebeam is still changing. At times, that has led to frustration because the process has never been consistent from project to project, but it has equally been rewarding, each time getting a little better, a little more valuable, or a little faster. He does believe Black & Veatch is reaching the plateau of a consistent, refined review process, but he also notes, "You never know what Bluebeam will introduce next."

Looking back, Larry cites the importance of allowing the organic growth of a software package, never giving people an ultimatum or forcing them into a corner. He believes if you give people access to a good product and teach them how to use it in a good way, they will make the right decision and do great things with that product.

Larry can't imagine a project without Revu, and he's happy the business has recognized its benefits so he no longer has to fight for it. In fact, the business has even recognized the value of those larger monitors Larry promised his team, so he no longer has to use them as a bargaining chip.

When it comes time for your company to make a change, think of Larry. Even in an enterprise of more than 10,000 professionals, with the right drive and some perseverance, a single person can make a big splash. Today Black & Veatch has over 5000 licenses of Revu and uses the software in each of its business units. It has a dedicated user expert team to help resolve Bluebeam problems, an excellent relationship with Revu developers, has hosted a Bluebeam Hack-a-Thon, and annually holds a 5-day Revu training event called Bluebeam Week.

For IT

Now, once a business has decided to invest in Bluebeam, there are some back-office IT tips to be aware of. Bluebeam stores user licenses on its own servers, not on the servers of its customers. This means that users need to be connected to the internet in order to install the software and to receive version updates. For most regular users this is not a problem. For users consistently not connected to the internet, for example in sensitive compartmented information facilities where users work with classified information and have no network or internet access, the license authorization process can be done manually.

1. Under the Help tab in Bluebeam Revu, click the Register button.
2. Enter the serial number and product key and click Register.
3. Click Get Authorization Code Manually, enter the computer name, and then click Continue.
4. Copy down the serial number, product key, and security ID.
5. From a machine with Internet access, open a web browser and browse to www .bluebeam.com/authorize.asp.
6. Key in the serial number, product key, security ID, and computer name.
7. Click Get Authorization Code. The authorization code will display on the next page.
8. Copy down the authorization code and return to the computer where the Bluebeam software is being installed. Enter the authorization code into the Manual Authorization window and click Authorize.
9. Authorization is complete.

Bluebeam is generous with license overages for enterprise customers, allowing an extra percentage of the total paid license count and then "truing up" the additional cost with the customer. This releases IT from saying no to users who urgently need Bluebeam, even if the company is maxed out on its license count. Bluebeam is also understanding for large enterprises whose users may not all upgrade versions at the same time, or who constantly have computers retiring. Each version of Bluebeam Revu installed and in use on a computer requires its own license, so if a user is in the upgrade process from version 2015 to 2016, he or she will temporarily have two licenses occupied on the single machine. Bluebeam does not count the temporary older license against the total enterprise seat allowance, and even allows an enterprise to have some seats in the newer version and some in the older version to account for a gradual upgrade process. For retired computers where Bluebeam Revu is not uninstalled before the computer goes away, there is a wait- ing period of two weeks where the license is still counted as in use, and then it is freed up. For more information on enterprise administration of Bluebeam, visit http://support .bluebeam.com/enterprise-installation/. Bluebeam sends monthly license usage (Table 1-3) reports detailing the number of licenses allowed, the number of licenses in use, and other helpful IT management information.

In addition to comparing the base license cost of Bluebeam versus Adobe, a company should also consider the overall cost of transitioning from red to blue. Anecdotally, Bluebeam initially costs more in IT service ticket spend because it is new to most of the end users and because the end users are utilizing Bluebeam in deeper, more complicated ways than they typically used Adobe. Once users become familiar with Bluebeam, the service ticket count decreases and levels off close to the old Adobe service ticket count. IT may spend more overall time packaging Bluebeam updates for deployment than Adobe, because Bluebeam releases version updates with bug fixes a few times per year, and Adobe historically released version updates only every three years. It is up to the particular company how frequently they update the version. Especially in companies with a wide range of users, from Bluebeam groupies to old-school paper devotees, IT will never make everyone happy with the frequency of updates. A company should find an update rhythm that works for its users; it can always deploy an

Table 1-3: Sample Monthly License Usage Report

Serial Number	Product Key	Version	Maintenance?	Maintenance Exp. Date	Registered Email	Seats Allowed	Seats Installed
1237894	A5GS9–7JCK1L3	Bluebeam Revu 2016 eXtreme	Y	5/14/2017	ITmanager@ company.com	500	398
6540983	I6ZL6– 4QKF9W1	Bluebeam Revu 2015 eXtreme	Y	5/14/2017	ITmanager@ company.com	400	95
4561238	MXY4F– BHE8UP5	Bluebeam Revu 12 eXtreme	Y	5/14/2016	ITmanager@ company.com	0	4
					TOTAL:	**900**	**497**

emergency update if needed for a certain bug fix or needed feature. Training will be needed for new users of Bluebeam, which takes time. Bluebeam themselves as well as several registered resellers offer training for a fee. If a company is lucky enough to have motivated and generous Bluebeam enthusiasts, it's often best to harness internal resources to create targeted training courses with familiar example material. Likely, some processes and procedures will need to be rewritten when a company switches from red to blue. For example, a company has a procedure with specific instructions on how to create and apply digital signatures in Adobe. The end result is the same in Bluebeam, but the specific steps are different. Modified procedures will need to be approved and socialized. To summarize, some of the main contributing line items to the overall transition cost from red to blue are:

- Familiarizing IT with different installation and license management processes
- Handling increased service desk tickets for the initial adjustment period
- More frequent software version updates
- Training for end users
- Modification of Adobe-specific procedures

Training

For the end user, the adjustment from red to blue is varied. For a tech-savvy user, the new interface is intuitive and exciting, letting him or her do much more than they could with Adobe. For users who don't easily understand technology to start with, switching over to a new program feels like cruel and unusual punishment. Chapter 2, "Doing Red in Blue," will explain how to do typical PDF functions in Bluebeam. In this chapter, it's necessary to discuss the importance of training for new users. It takes a concerted effort to successfully roll out new technology. There is the initial value proposition to create, then the convincing of management, then the actual IT steps to procure and install the software, then the change management and training for end users. If any of these actions are neglected, the software has a good chance of not making it, rendering the initial value proposition useless.

A company may choose to pay for external training, or may empower internal trainers. Experience says that internal trainers are cheaper, more relevant to their particular company's business, and available to help even after the training class is over. The keys to a successful internal training program are passion, discovery, support, and generosity. The trainers need to care about educating their peers. They should "believe" in the software. Having passion for a cause fuels people, and they will go to remarkable lengths above and beyond the nine-to-five call of duty when they feel what they're doing is right and necessary. The trainers need to be discovered, by management and by each other. In a large organization, passionate trainers may be completely siloed from one another. It takes a manager with a broad and deep gaze to identify people across the business and at perhaps low levels who would be good trainers. Then that manager should bring them together. Let them identify with each other and swap ideas and experiences. The only thing more powerful than a passionate person is a united group of passionate people.

The manager who is sponsoring the transition from red to blue doesn't have to become an expert in the software to give it a successful rollout (what upper manager has time for that?). But that manager does have a responsibility to raise up leaders who can shepherd their fellow users through the transition. After the trainers meet each other and understand their charter, management needs to support their work. Most professionals are already overly busy with their day jobs; adding the responsibility of creating training content, leading training sessions, and spending one-on-one time answering users' questions afterward can take up a surprising amount of time. Managers need to be understanding of these responsibilities and afford the trainers the time they need to make the transition successful. Training should be counted as a high-value activity, because the impact of one person's class ripples through dozens, maybe thousands, of employees, which saves an exponential amount of time when compared to letting each user fend for him- or herself.

Finally, generosity on the part of the trainers is needed. Generosity of time, patience, and listening are key to being a successful trainer. Many times, the trainer will encounter users who are quick to point out mistakes or like to stump the trainer. Respond graciously, write down the question and the name of asker, and follow up. Other times, the trainer will answer a phone call with the same basic question for the fifth time that day. Answer anyway, instruct gently and patiently, and leave the user with a way to answer his or her own question in the future (e.g., the Bluebeam Help menu or Google). Being the expert at Bluebeam—a software that touches almost everyone in an architecture, engineering, construction firm—is a great way to meet new people and serve them, creating a new network where the professionals served leave with an outstandingly positive impression of the trainer's attitude and intelligence. Remember, serving faithfully in the small ways recommends a person for bigger responsibility.

Some practical ideas for training include:

- Spread the training out among several individuals; don't leave one person to shoulder the burden alone. Give each trainer responsibility for teaching his or her favorite features of Bluebeam.

- Corral as many people as possible into the basics sessions. Most everyone needs to go through this class when first using Bluebeam, and it gets exhausting to keep training small groups of five or six people on the same thing.

- Make a big push for Bluebeam training once a year instead of stringing out low-visibility courses throughout the year. Partner with managers, administrative assistants, and IT to get the word out about upcoming training and reach as many people as possible in each session.

- Create different training courses for different types of users. Administrative assistants and engineers need to know different things in Bluebeam. Don't make them sit through each other's classes.

- Record, record, record!! Invariably, after a week of hearty training, a brand new professional emails to say he or she just started at the company and wonders about getting an individual make-up session. No. Direct these people to an intranet site where the recordings are posted and clearly labeled.

- Create an online resource center, maybe on SharePoint, with any guides, instructions, recordings, FAQs, shared tool sets, and so on. Tell every person who is trained about this site and email out the link often. The best way to train professionals is to empower them to learn of their own free will.

- Recognize potential new trainers. Grow the circle of trainers by recognizing skilled, curious users with good communication skills. Suggest they help out with a future training and give them a piece to teach.

- Protect Bluebeam Support if possible by taking care of simple questions within the company. Short story: A user once claimed he absolutely could not login to Bluebeam Studio; he had tried everything. A trainer assumed the issue was serious and engaged Bluebeam Support to fix the problem. One long email chain later, it was discovered that the user did *not* have Bluebeam Revu installed at all.

EXPERT TIP

Setting the Default PDF Application

Many professionals do not know how to change the default program used with certain file types, such as PDF. At one large organization, it was decided to swap Bluebeam Revu for Adobe Acrobat, but leave Adobe Reader so that users had a backup PDF viewer. This left users befuddled as to how to open a PDF in Bluebeam Revu instead of Adobe Reader, besides right-clicking to open in a different program each time they opened a file. To change which program operates with PDFs by default (in a Windows operating system), right-click on any PDF in Windows Explorer or on the desktop. Hover the mouse over Open With and click Choose Default Program. A list of recommended programs will appear in the Open With dialog box. Click on Bluebeam Revu, make sure the "Always use the selected program to open this kind of file" is checked, then click OK. The user will notice that all PDF file icons change from the red Adobe icon to a blue Revu icon, indicating that the switch was successful.

Conclusion

To recap, the transition from red to blue typically follows these steps:

1. Determine the use case for Bluebeam. How will it provide more needed functionality than Adobe?
2. Create the value proposition. How much time and money will be gained if the company switches PDF software?
3. Get management buy-in and budget.
4. Prepare IT.
5. Start switching users in groups from red to blue.
6. Identify and gather expert users to become trainers.
7. Create and deliver training content; build a content library.
8. Modify and socialize PDF-software-specific procedures.
9. Monitor license count and support end users.
10. Reap the rewards! Record lessons learned and iterate through increasingly efficient uses of Bluebeam.

Chapter 2
Doing Red in Blue

When a PDF-inhaling user takes the leap from Adobe to Bluebeam, the hardest part is not getting used to the very AEC-specific tools Bluebeam Revu offers. It's figuring out how to do those baseline PDF functions that were so easy in red—because it was familiar—which are suddenly hidden in the hieroglyphics of a new user interface with new icons and new button locations. This is daunting for any user and causes what is dubbed, "the three days of hating Bluebeam." The first three days for a new Bluebeam user can be frustrating. Kind of like when a person gets married and the last name changes. The change is good, the new name was willingly accepted, but it is still hard to get over the heuristics of the way things used to be, at least for the first few days. Bluebeam does offer enough workhorse PDF functions to invite the user into a monogamously blue relationship, and an acceptance of the short learning curve will make the transition much less painful.

Changing Preferences

To start, the first time Bluebeam is opened, the new user will realize the default color scheme is black and blue rather than white and red. If this is bothersome, change the color scheme to be more comfortable before taking on any other functions. To change the color scheme, first click on any of the tabs at the top of the screen, then click on the Settings button in the upper right hand corner of the screen, and click on Preferences (Figure 2-1). Ctrl+K will also open the Preferences dialog window.

By default, the General tab will be open. Find the Brightness meter toward the bottom of the window and adjust it until the color scheme hits the desired lightness. Checking the Classic Mode box changes the user interface to look similar to Adobe's interface, with a white background and small tabs.

Stay in the Preferences window a while longer, and click on the Navigation tab (Figure 2-2). The Mouse Wheel section of this menu allows the user to define how the mouse center scroll wheel works in different situations. There are two main view modes in Bluebeam Revu: Single Page Mode and Continuous Mode. In Single Page Mode, only one page will appear at a time, and the user has to switch pages by using arrows at the bottom of the interface. In Continuous Mode, multiple pages appear on the screen at a time, and the user can switch pages by scrolling up and down. One of the most frustrating things for new users is when they open a multi-page PDF and they can only zoom in and out on the first page. This happens because the document opens up in Single Page Mode; if they switched to Continuous Mode,

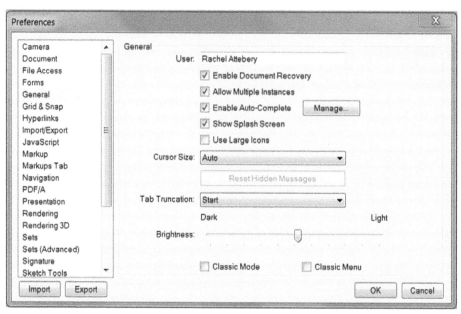

Figure 2-1: Preferences Window

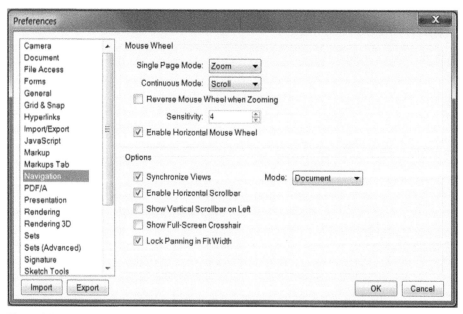

Figure 2-2: Scroll Wheel Behavior Options

the scroll wheel would roll between pages rather than zooming in and out. Nevertheless, users can configure the scroll wheel's behavior in these two modes to suit themselves. To change the scroll wheel's behavior for a certain mode, click on the Zoom/Scroll dropdown menus and make a selection. Setting Scroll for both modes allows the user to use the scroll wheel to switch pages in a document regardless of in what mode the document opens.

The third Preference tab that will bring a new user relief is the File Access tab. In this tab, integrations with document management systems are controlled. Bluebeam has built-in integrations with Bentley's ProjectWise and Microsoft's SharePoint. More information about these integrations can be found in chapter 6. Bluebeam can automatically detect the path to any SharePoint document that is opened in Bluebeam, and it can automatically set up an integration with that document library. Sometimes this is helpful, because the user can open and check in and out files directly between SharePoint and Bluebeam without downloading any files to the desktop first. However, sometimes this is annoying, because the user is working out of a different SharePoint path, or a different document management system, or just his or her desktop, and Bluebeam is indefatigable in giving the user multiple options on where to save the working document. To turn off the automatic integration of SharePoint sites, click on SharePoint Options (Figure 2-3) and uncheck "Auto detect SharePoint network paths."

The Preferences window is full of helpful options that the user can customize to make the best possible individually tailored Bluebeam experience.

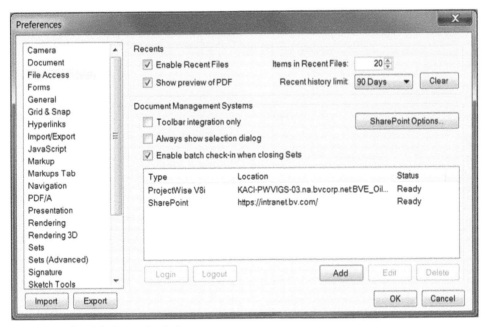

Figure 2-3: SharePoint Integration Options

Tabs and Toolbars

Another big help for first-time users is customizing the visible toolbars to pare down the number of buttons pinned to the home screen. Bluebeam Revu can do a lot, but most users don't need shortcut buttons to every single function available. To change what tabs and toolbars are visible by default, go to the View tab and peruse the tabs and toolbars buttons. Tabs appear in the left, right, and bottom flyouts. Toolbars are pinned to the top, bottom, right, or left of the screen.

Available tabs are shown in Figure 2-4.

Available toolbars are shown in Figure 2-5.

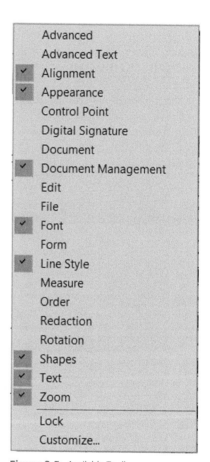

Bookmarks	Alt+B
File Access	Alt+A
Forms	Alt+Q
JavaScript Console	Alt+J
Layers	Alt+Y
Links	Alt+N
Markups	Alt+L
Measurements	Alt+U
Properties	Alt+P
Search	Alt+1
Sets	Alt+2
Signatures	Alt+4
Spaces	Alt+S
Studio	Alt+C
Thumbnails	Alt+T
Tool Chest	Alt+X
3D Model Tree	Alt+3

Figure 2-4: Available Tabs

Advanced
Advanced Text
✓ Alignment
✓ Appearance
Control Point
Digital Signature
Document
✓ Document Management
Edit
File
✓ Font
Form
✓ Line Style
Measure
Order
Redaction
Rotation
✓ Shapes
✓ Text
✓ Zoom
Lock
Customize...

Figure 2-5: Available Toolbars

A highlight reel of the toolbars for basic PDF actions:

Advanced: Hyperlinks, Stamps, Attachments

Advanced Text: Edit Text, Review Text, Underline/Squiggly/Strikethrough Text

Alignment: Align, Distribute, Flip Objects

Appearance: Object Outline Color, Fill Color, Transparency, Hatch

Document: Headers & Footers, Crop Pages, Flatten, Erase Content

Edit: Undo/Redo, Cut, Copy, Paste, Format Painter, Delete, Snapshot, Flatten, Erase Content

Font: Change Font, Font Size, Font Color, Bold/Italics/Underline, Left/Center/Right Alignment, Superscript/Subscript

Order: Bring to Front, Send to Back, Bring Forward/Backward

Redaction: Mark for Redaction, Apply Redactions

Rotation: Rotate Page Counterclockwise/Clockwise

Shapes: Draw a Line, Arrow, Callout, Rectangle, Oval, Polygon, Insert Image, Crop Image

Text: Insert Text Box, Typewriter, Sticky Note, Flag, Pen, Highlighter, Eraser

To further customize toolbars and which actions are pinned to each one, click Customize at the bottom of the Toolbars button menu. All toolbars can be moved around the screen and docked wherever the user wants by grabbing the four vertical dots on the left side of the toolbar. Likewise, tabs can be moved among flyouts by grabbing the tab icon and dragging it to the desired location.

Creating PDFs

Now that some of the biggest potential annoyances for a new user are addressed, it's time to get into the workhorse PDF functions. In order to do anything with a PDF, there first must be a PDF. There are three main ways to create a PDF from some other native file type.

1. Print to Bluebeam PDF is found along with other physical printer options in the print dialog of most programs.

2. Save As PDF is a regular Save As function for a document, but changes the file type to PDF before completing the save.

3. Bluebeam Plugin Create PDF was created by Bluebeam to integrate into many programs, such as Microsoft Word, Excel, PowerPoint, and Outlook; Autodesk AutoCAD, Revit, and Navisworks; SolidWorks and Sketchup. The plugin automatically installs into these programs when Bluebeam Revu is installed on a user's computer; refer to chapter 1 to see which versions of Bluebeam Revu have a plugin for which programs.

EXPERT TIP

PDF Creation

For most generic PDF creation cases, any of these three PDF creation options is fine. However, there is a difference in results from each option that certain users need to be aware of. For example, the Bluebeam Plugin transfers active hyperlinks and bookmarks from the native file to the PDF, while Save As PDF transfers hyperlinks but not bookmarks, and Print to PDF transfers neither hyperlinks nor bookmarks. The Bluebeam Plugin sometimes has trouble correctly transferring formatting and photos and symbols, while Save As PDF creates a perfect PDF replication of the native file. Save As PDF can be considered the best option for marketing and sales materials, which are photo- and formatting-heavy. The Bluebeam Plugin can be considered the best option for text-only documents with a complex bookmark hierarchy.

The user can also create PDFs from within Bluebeam Revu by using the Create button under the File tab. There are five options for PDF creation under this button.

1. From File: navigate to a single file and turn it into a PDF

2. From Multiple Files: navigate to multiple files of differing types and turn them into either one combined PDF or separate PDFs

3. From Scanner or Camera: select a connected scanner or camera device and convert the content directly from the device into a PDF

4. Layered PDF: navigate to multiple files and flatten them on top of each other into one PDF

5. PDF Package: navigate to multiple files of differing types and add them to the PDF Package, which can be thought of as a file container, but is still a PDF file type

When creating a PDF from multiple files, the Bluebeam Stapler application will open (Figure 2-6). Stapler can be accessed outside of Bluebeam and has the sole function of converting files to PDFs and combining them together.

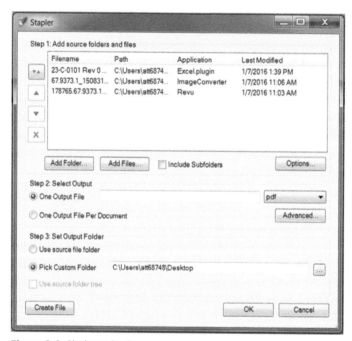

Figure 2-6: Bluebeam Stapler

With Stapler, multiple files and even folders of files can be queued for conversion to PDF. Choose if the resulting PDF should be combined into one, or kept as separate PDFs. Choose a destination for the resulting file(s) and click Create File to start the process. Stapler offers the useful feature of being able to save a "job." For example, many AEC documents go through versions. Different people on a given project need to see the same documents many times as the documents develop through the course of a project. Often, many different files are combined into one PDF report, and the same report will be submitted for review with different versions of the files throughout the lifecycle of the project. Once a Stapler job is set up with the desired file paths and settings, it can be saved and automatically run at any time in the future without having to renavigate to the files or redo the settings. To do this, click OK instead of Create File in the Stapler dialog window. Another Stapler window will open (Figure 2-7). Click File, Save, to save the job as a .bsx, Bluebeam Stapler, file. Anytime the job needs to be rerun, just open the .bsx file and click Staple. The Stapler job can even point to files that exist in SharePoint or ProjectWise; so imagine the possibility of quickly creating an equipment specification PDF from the latest versions of ProjectWise documents without having to renavigate to each file first.

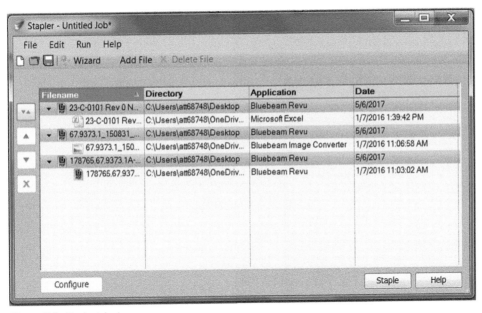

Figure 2-7: Stapler Job, .bsx

PDF Document Actions

Combine PDFs

A user can combine multiple PDFs into one by using the Combine button under the File tab. Conveniently, the Add Open Files button adds all files currently open in Bluebeam to the queue. There are options for what to carry over into the combined PDF: include bookmarks, include file attachments, merge document properties, merge layers, and use filename as page label. After a PDF is created and open in Bluebeam, all the typical modification functions can be found under one button: Document tab, Pages button (Figure 2-8).

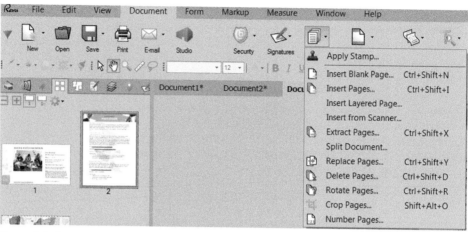

Figure 2-8: Document Pages Actions

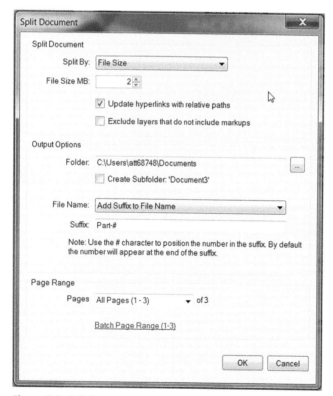

Figure 2-9: Split Document

The following functions live here:

Apply Stamp

Insert Blank Page

Insert Pages

Insert Layered Page

Insert from Scanner

Extract Pages

Split Document

Replace Pages

Delete Pages

Rotate Pages

Crop Pages

Number Pages

Most of these functions are straightforward. One to call out is Split Document (Figure 2-9).

Split Document

Split Document is useful when a file is too big to send via email or to upload. The document can be split by page size or file size. So, for instance, if the original file is 20 MB and it is split by file size into 2 MB chunks, it will be split into 10 separate files, each 2 MB large. Use the file name options to add a prefix or suffix to the separate files, such as "Part 1." The prefix and suffix can be customized to say, for example, "Part 1 of 10." The Page Range option at the bottom of the Split Document window allows the user to specify which pages in the document get split.

Replace Pages

Replace Pages is very useful for swapping out old or incorrect pages with the correct page (Figure 2-10).

Multiple pages can be replaced by specifying the Remove Pages range and can be replaced by multiple pages by specifying the Replace With range. The "Replace page content only" checkbox will retain information like bookmarks, markups, and page labels and literally just replace all page content while protecting the properties of the page itself. "Replace

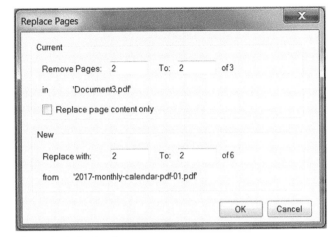

Figure 2-10: Replace Pages

page content only" can be thought of as slipping a new layer of content under any imposed markups or properties, and deleting the old layer of content. This is helpful in a situation where markups need to stay intact, such as when backchecking redmarks. The new version of a drawing can be slipped under the markups and easily checked for accuracy.

Reduce File Size

Besides the document manipulation functions that exist under the Pages button, there are several other features commonly used in red. A big one is Reduce File Size (Figure 2-11). This function lives under the Process button in the Document tab.

Figure 2-11: Reduce File Size

Bluebeam will analyze the document and provide an estimate of how much it can reduce the file using the default reduction settings. The reduction settings are balanced between quality and compression. Move the slider bar over toward quality to retain better-looking images, sacrificing the amount the file can be reduced; or move the slider bar over toward compression to end up with a smaller file, sacrificing image clarity. Sometimes, the user may want to get more granular with the file reduction settings. To use custom settings, click the Custom radio button and then click Edit.

The Reduce File Size Custom Settings (Figure 2-12) shows an analysis of the file's content at the bottom of the window. This helps the user to know where most of the file's size is coming from. In this window, the user can specify how to treat images. For example, perhaps it is fine to reduce the quality of some grayscale images, but the full-color images need to retain their full quality. The user can find the Full Color option under the Images tab and select No Change instead of the default JPEG. The Max DPI, Bit Depth, and overall Quality of the image can also be specified. There are options under the Fonts and Miscellaneous tab as well that can help the user pick and choose which elements get dropped or decreased in quality during the file size reduction process. If the user expects to reuse these settings, he or she can save them by clicking Save at the top of the window and giving this setting set a name. In the future, when the user starts the Reduce File Size dialog, he or she can choose the saved settings from the dropdown options next to the Custom radio button.

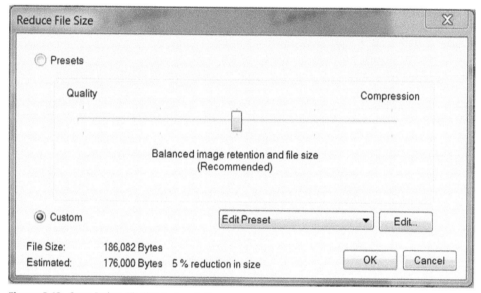

Figure 2-12: Open Reduce File Size Custom Settings

EXPERT TIP

Reduce File Size

Many times, the reduce file size estimate is conservative. To find out exactly how much the file will be reduced, go ahead and run the reduction process. Often, the final file will be smaller than was predicted.

Bookmarks and Page Labels

Bluebeam has the ability to create bookmarks and page labels. While both can be created manually, there is a fantastic feature called AutoMark that allows the user to select a region of the page and use Optical Character Recognition to pull that text and turn it into the bookmark or page label automatically. Once either the bookmarks or page labels are created, they can be copied over to each other. So, if the user creates a set of bookmarks, he or she can easily copy those over to be the page labels as well. The Bookmarks tab is found by default in the left-hand flyout and looks like a little blue ribbon (Figure 2-13). If the bookmarks tab is not shown, right-click in the blank space of the flyout's toolbar, click Show, and turn on the bookmark tab.

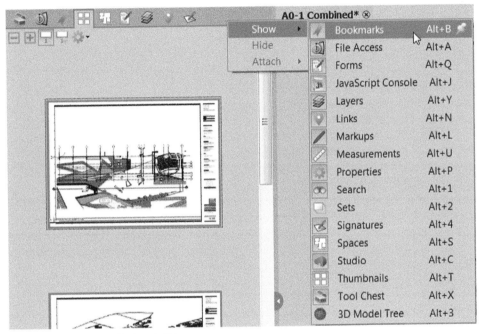

Figure 2-13: Turning on Bookmarks Tab

For a combined PDF that has been created from other files, the names of the original documents will automatically be created into bookmarks for the different sections of the combined file. For Excel documents, the different tabs of a worksheet are made into bookmarks as "Child" bookmarks to the file name. To create bookmarks, open the Bookmarks tab. Click on the icon with a blue ribbon and orange star (Figure 2-14).

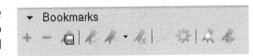

Figure 2-14: Create Bookmark

This will open the Create Bookmarks dialog window. If the document already has page labels created, the user can select the Page Labels option to transfer the page labels over to the bookmarks. If starting from scratch, click the Page Region option and click the Select button. The mouse cursor will turn into a crosshairs, allowing the user to select a region of page to pull text from, which will become the bookmark. Whatever region is selected will be used for all pages of the document, unless the user specifies the page range. Zoom in on

sheet number

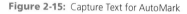

Figure 2-15: Capture Text for AutoMark

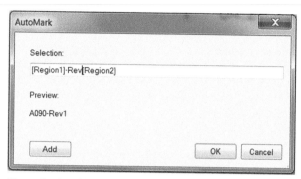

Figure 2-16: Add Regions to AutoMark

the page region to pull text from. This may be part of a header or footer, or for engineering documents, wherever the sheet number is located (Figure 2-15). Use the mouse to draw a box around the entire area where the text appears. Make sure to draw the box big enough to capture all the text on each page, since some text will be longer than others from page to page.

The AutoMark dialog window will appear and show a preview of the text captured. If this text is correct, click OK to pull text from every page and create the bookmarks. If it is not correct, click Cancel and redraw the capture area. Sometimes it is useful to concatenate more than one region of text into a single bookmark. Perhaps it is important to capture both the sheet number and the revision. To do this, click Add in the AutoMark dialog window and select another region of text (Figure 2-16). The two text regions will be combined to create a multi-data bookmark. Add a divider, like "-" or a space between the two pieces of text by clicking in the Selection line and adding any additional text.

Click OK to create the bookmarks. Check the bookmarks and make sure all text was pulled correctly. Sometimes there will be a mistake, and that can be corrected by right-clicking on the bookmark and selecting Rename. Edit the bookmark and press Enter. A bookmark can also be created by navigating to the page needing a bookmark and clicking the icon with a blue ribbon and green plus sign.

The user can then manually type in the name of the bookmark. The dropdown menu next to this icon allows the user to add bookmarks before or after the selected bookmark, as well as add a Child to the selected bookmark. The bookmark hierarchy can get quite complex. For some documents, the user may want to set and save a certain collapse state of the bookmarks, meaning that when someone else opens up the document and uses the bookmarks to navigate, certain bookmarks are showing and others are collapsed. To save the collapse state, click the icon with a blue ribbon and the save disk. Other bookmark actions include deleting bookmarks, editing the action of a bookmark, changing its appearance, auditing the bookmarks to make sure none are broken, and exporting the bookmarks to create a table of contents.

Edit Action: The default action for a bookmark is to jump to the page for which the bookmark was created. Essentially, it is a hyperlink. The user can edit the bookmark's action to jump to another place in the document, zoom in on a certain section or image, open up a webpage, or even open another document.

Properties: The bookmark's title, color, and font boldness or italics can be edited. This is helpful for highlighting certain bookmarks within a hierarchy.

Creating a table of contents is a valuable ability, and though Bluebeam doesn't offer a simple button to create a table of contents, it is possible to make one using bookmarks and page labels. Check out this article for the best way to create a table of contents. http://support .bluebeam.com/articles/how-do-i-create-a-table-of-contents-in-revu/

The Export Bookmarks button offers a rudimentary way to create a table of contents by exporting the bookmarks to PDF (or CSV) and creating a hyperlink between the bookmark titles on the new PDF and the pages from the original PDF. Add in the new PDF at the beginning of the original document to provide an easy way to navigate through the pages without using the Bookmarks tab.

Once the bookmarks are created, it is quick and easy to create page labels. The reverse is also true; whether the user starts with bookmarks or page labels, either can be copied over to the other. Following along with the example of creating the bookmarks first, jump over to the Thumbnails tab in the left-hand flyout (left-hand by default, or wherever the user has moved it). The Create Page Labels button looks like the bottom of a white page with a yellow star (Figure 2-17).

Figure 2-17: Create Page Labels

Choose the Bookmarks radio button to copy the bookmarks over to the page labels, or select Page Region to select a new region of text as the page labels. Use the Page Range to select which pages will be labeled this way. Click OK and watch the page labels appear!

Meet Susan Taylor

Office Manager for the Chairman & CEO, Black & Veatch

Photo Courtesy of Susan Taylor

"My job is to manage the office of the chairman and CEO. Part of my job is to create the board book five times a year for the board of directors to review for board meetings. The board meetings are very important, so the books have to be right. I use Bluebeam to do this now. I've adapted to it and really like it. At first it was scary to switch from Adobe to Bluebeam, just because it was something new, but I thought, 'I just need to do it because it will save the company money,' and I was going to have to switch eventually anyway. I didn't switch until after the board meeting, so I had time to play with it! It wasn't that bad, and there are things I like better in Bluebeam. The training provided was extremely helpful in making the transition. I have a printout of the training presentation, so I can refer back to it when I get stuck. It seems like I had trouble with the page numbers until I figured out how to do it. I was leery of pulling new pages into the document because it can be 200–300 pages long, and if they drop into the wrong spot, it can be hard to figure out what goes where, but with Bluebeam it is much easier. It is also easier to create PDFs from Word and Excel files with Bluebeam because the Create PDF icon just does it with one click. With Adobe, I always did File, Save As PDF, and it took a little more time. It's simpler with Bluebeam and was easy to figure out.

"It took no time at all to figure out how to do my daily activities in Bluebeam, since opening PDFs is easy. If someone wants me to replace pages in a document, I go to the Thumbnails tab so I can see all the pages and select the starting and ending pages of the section I'm replacing. I'm at work late at night sometimes putting together the board book; and when you're tired at night, it helps a lot to see where you're at. I got rid of all the toolsets I don't need, so I wouldn't be distracted by buttons I don't use. I love that inserting pages and rotating pages is easier with Bluebeam. I can just select a page or group of pages and rotate them with a click of a button, but in Adobe sometimes it would rotate the entire document. I use the bookmarks all the time and learned how to format them so certain sections stand out. I'm still learning the features, but after I do something the first time, it goes a lot better the second time.

"The directors like the electronic copy of the board book because they have it with them on their computers at all times, and they don't have to carry around the big heavy paper version. It's really not that hard once you switch over. I think it's just a matter of doing it. The worst part was the scariness of something new, because it wasn't too bad once I started doing it. My advice to new users is, 'You can do it!'"

Forms

Premade form fields can be filled in Bluebeam Revu the same as in Adobe Acrobat. Simply click in the field to type text or select an option. Creating forms is an important and advanced activity and is covered in chapter 9, "Go Digital, Document Assembly."

Headers and Footers

To add headers and footers, jump again to the Documents tab and click on the Headers and Footers button. The Header and Footer dialog window will appear, giving options to add headers and footers on the left, center, or right side of the document (Figure 2-18).
The font, text size, text color, and text emphasis (bold and italics) can be changed. In addition to any text added manually, certain document properties can be added automatically, such as Page Number, Date, Bates Number, and File Data. The headers and footers can be positioned on the document by adjusting the margins. Use the Batch Page Range link in the bottom right-hand corner of the window to add other documents to apply these headers and footers to, and to adjust the page range where these headers and footers are applied. Using the Page Filter options in this window, the headers and footers can be specified to apply only to even- or odd-numbered pages or to only landscape- or portrait-oriented pages.

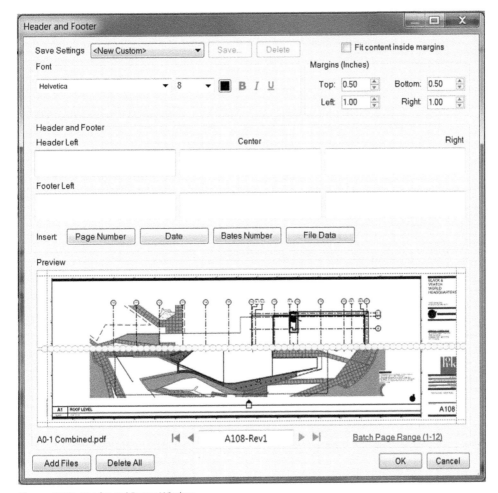

Figure 2-18: Header and Footer Window

This is helpful for multi-page documents which might have pages of different orientations that would end up with the header and footer in an awkward position on the page. Save headers and footers for reuse by clicking Save at the top of the Header and Footer window. Type a name for these settings and click OK. In the future, reapply these saved settings by selecting this set from the Save Settings dropdown menu at the top of the Header and Footer window. At this time, images cannot be added to headers and footers through this dialog window. However, an image can be placed on the document, positioned, and then applied to all pages. This will apply the image to every page in the document in the same position. To apply the image to a certain range of pages, pop out the Thumbnails tab and use the Ctrl key to select the pages where the image should be applied. Then right-click on the image and select Apply to Selected Pages, and the image will only be applied to the pages selected from the Thumbnails tab.

Color Processing

Color Processing enables the entire color of a document to be modified. To do this, click on the Process button under the Document tab. Click Color Processing. There are several Process Type options:

- Modify Colors: Allows a source color in the document to be swapped out for a different color (Figure 2-19). Multiple source colors can be changed at once. Tolerance can be adjusted to pick up on shades of a source color and turn all the shades to a single new color. Modify Colors only works with vector content, not raster content.

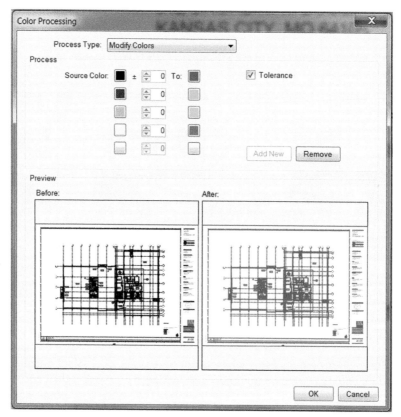

Figure 2-19: Modify Colors

- Colorize: Allows an entire document to be recolored with a range of colors (Figure 2-20). Use the Select Color option to set the darkest color, and use the To option to set the lightest color. The Scale Color checkbox will allow a range of colors within the set colors; unchecking the Scale Color checkbox will convert the content to one color only. Colorize only affects vector content, unless the Process Images option is checked.

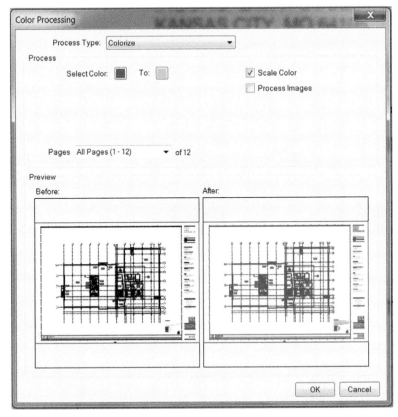

Figure 2-20: Colorize

- Grayscale: Converts colors to grayscale. Like Colorize, Grayscale affects only vector content, unless the Process Images option is checked.

- Black and White: Converts colors to black and white. Again, Black and White affects only vector content, unless the Process Images option is checked.

- Luminosity, Saturation, and Hue: Familiar image processing options that allow tweaking of the document content. Affects only vector content, unless Process Images option is checked.

- Mask Images: An extremely useful option for removing the background of a document or image (Figure 2-21). Choose a Mask Color, the color that will be removed, and adjust the Tolerance to remove as much of that color as desired.

Redact and Edit

Sometimes it's important to be able to redact, edit, or cut content on a PDF. These features live under the Edit tab. To redact content, click the Redaction button and then click Mark for Redaction. Bluebeam will give a warning explaining what redaction is and that it

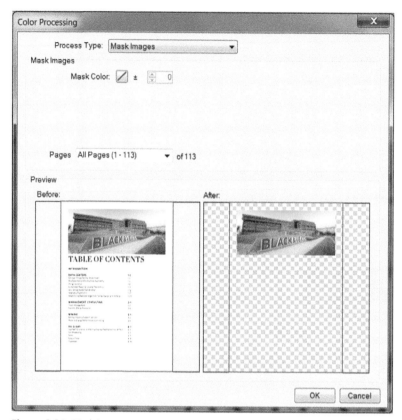

Figure 2-21: Mask Images

is permanent. Click OK to continue, and use the mouse cursor to select areas of the PDF to be redacted. The area marked for redaction will be surrounded by a black box if it is purely graphical, and with a black and red box if it contains text. After all areas to be redacted are marked in this way, go back to the Redaction button and click Apply Redactions. Bluebeam will show another dialog window explaining that the redaction is not permanent until the file is saved. It will also provide a list of additional document content that can be redacted, such as bookmarks, markups, metadata, hyperlinks, document properties, and form fields. Choose the appropriate options and click OK. The redacted areas will appear as black boxes. Save the PDF to finalize the redaction.

To edit or cut content without completely redacting it, use the Content button under the Edit tab. The three options are:

1. Edit Text: Edit existing text almost as if editing the native file. Change font type, text size, color, emphasis. Note that editing a PDF is not best practice for extensive format editing, and format should be edited whenever possible within the native file and then printed to PDF. Edit Text is best for small changes. To see edit text options, click Edit Text, open the right-hand flyout, and click on the Properties tab. Hit the ESC key to exit Edit Text mode.

2. Cut Content: Cut a section of content, whether text or graphics, and paste it elsewhere in the document. Use the mouse to draw a box around the content to be cut, then use right-click, Paste, or Ctrl+V to paste the content.

3. Erase Content: Cut content without copying it to the clipboard for repasting. Erase Content is similar to redaction but doesn't leave a black box behind indicating that content has been removed.

Search

Bluebeam has the ability to search a document both by text and by image. To search a document, open the right-hand flyout (by default) and click on the Search tab, which looks like a pair of green binoculars. If the Search tab does not appear in the right-hand flyout, right-click in the blank space of the flyout toolbar, click Show, and show the Search tab. Use the keyboard shortcut Ctrl+F to quickly open the Search tab (Figure 2-22).

In the Criteria section of the Search tab, there are two options: Text or Visual. The Text option allows the user to do a familiar search on key words or phrases. Notice the Search In dropdown menu, which allows the user to search for this word or phrase within the current document, the current page, all open documents, recent documents, an entire folder, or a current Studio Project. The Options section allows the user to specify which parts of the selected Search In option to look in (Figure 2-23).

Click on a result to jump to that location in the document. The checkboxes next to each result allow further actions. For any results that the user checks, the found word or phrase can

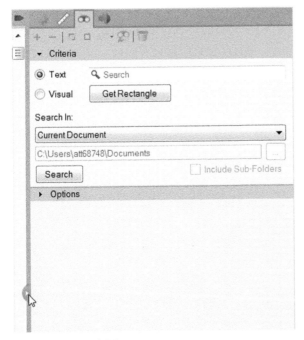

Figure 2-22: Search Tab

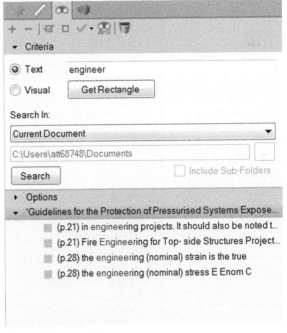

Figure 2-23: Search Options

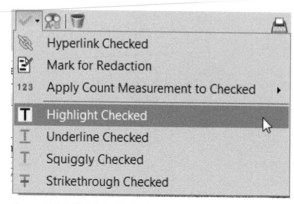

Figure 2-24: Highlight Checked Results

be hyperlinked, marked for redaction, counted, highlighted, underlined, underlined with squiggly, or strikethroughed (Figure 2-24). Use the Check All button to select all results at once, and use the Uncheck All button to uncheck all results at once.

The found word or phrase can also be replaced, similar to the search-and-replace feature in Microsoft Word. Use the Replace Checked button to do this. Click the Clear Search Results button, a trashcan icon, to start over.

Besides the typical text search, the user can search for a visual image. This is helpful for finding repeating images within a document and as a specific example, can be used to search a drawing for repeating symbols, such as globe valves, and then highlight or count the occurrences of those symbols in the document. To search for an image, click the Visual radio button, then click Get Rectangle (Figure 2-25). Similar to AutoMark, the user will use the mouse to draw a rectangle around the image to be found. In the Options for Visual Search, the user can adjust the sensitivity of the image search. Low sensitivity will give more leniency in the search and return more options. High sensitivity will perform a very stringent search and only return images

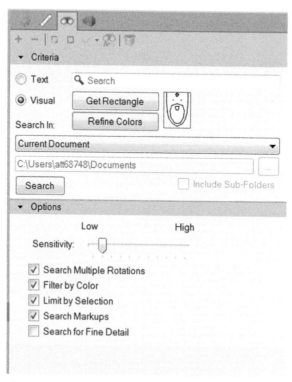

Figure 2-25: Visual Search

that look exactly like the selection. Also notice the options to search on Multiple Rotations, Color, Markups, and Fine Detail. It may take a couple tries to get the options just right to return the most accurate search results.

A Visual Search for toilets in a sample document with low sensitivity found all 181 toilets, even with slight variations in content (Figure 2-26).

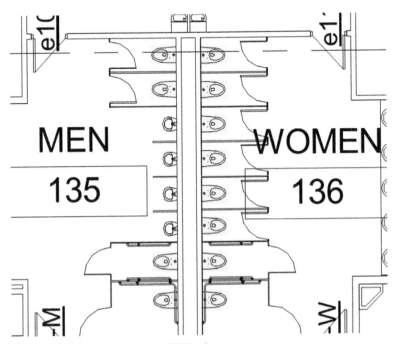

Figure 2-26: Visual Search Results Highlighted

Conclusion

Bluebeam Revu is a PDF workhorse that rivals and often surpasses Adobe Acrobat in its capabilities. This chapter covered the basic PDF functions that a new user needs to know to get off and running, and minimize the frustration of learning to do the same thing in a new program. The subsequent chapters will dive into more specialized features of Bluebeam Revu that go deeper than Adobe Acrobat's abilities.

Chapter 3
Redlining

Redlining or markup is the process of applying comments, notes, or annotations to a document. This could include any addition that is not native to the document file itself. For example, a reviewer may use a text box to suggest a change to the way a calculation is explained. In that case, the text box is a markup. Or, an engineer may sketch a new detail to be added to a construction drawing, in which case the detail is the markup. Redlining or markup can utilize anything from lines and text boxes to screen shots and graphics.

In the way of electronic redlines, it is not an exaggeration to say that Bluebeam revolutionized the industry. Markup or redlining a PDF is the single most powerful feature of Bluebeam, and further, in the author's opinion, it's the reason it has gained so much traction.

While many software platforms made electronic markups possible, it was Bluebeam that made them feasible. Until Bluebeam, the paper hard copy always won out on speed, efficiency, and ease of use. Redlining in other software platforms was clunky, frustrating, and time consuming. There was no question, paper was king.

Bluebeam overcomes the paper obstacle by enabling electronic redlines in a way that is easy, repeatable, and efficient. It contains a myriad of tool customizations, keyboard shortcuts that empower the user to utilize both hands, and tool sets that allow the simple reuse of commonly used marks. It also enhances the capabilities of the mouse and the user interface in ways that are unmatched.

This chapter focuses on how Bluebeam is suited to redline or markup. It will cover:

- Profiles
- Document navigation
- Toolbars
- Tool Chest
- Tool customizations

The User Interface

Diehard Bluebeam users are particularly fond of the User Interface (UI). Bluebeam developers have been able to make a UI that is not only functional, but highly customizable by the end user so each user is able to have a unique layout of the software, but still maintain all the power and functionality of Bluebeam. The UI can be as simple as that shown in Figure 3-1 or as complicated as that shown in Figure 3-2.

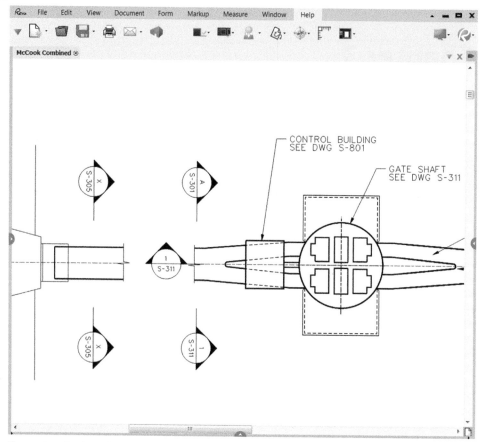

Figure 3-1: Simple Profile

Screen Layout

The Command Bar

The Command Bar, pictured in Figure 3-3, contains the various Revu menus such as File, View, and so on. Nearly every capability of Revu can be accessed through one of the menus on the Command Bar. Throughout the book, focus will be given to individual menus as they are needed for the various functions being discussed.

The Status Bar

The Status Bar is located on the right side of the bottom margin of the Revu window, as shown in Figure 3-4. In addition to indicating the size of the document, it can be utilized to toggle various behaviors and options inside the program.

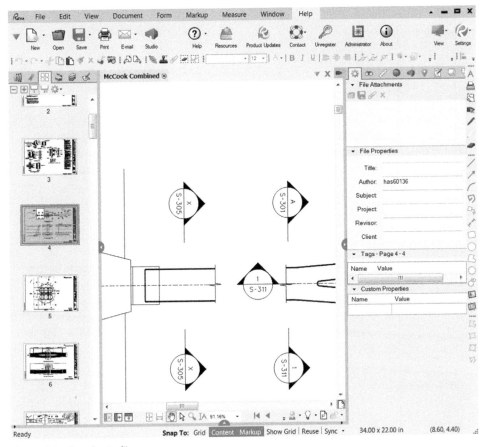

Figure 3-2: Complex Profile

Figure 3-3: Command Bar

Figure 3-4: Status Bar

The Navigation Bar

The Navigation Bar, shown in Figure 3-5, contains access to a number of settings that change the way the user navigates the document. The bar includes buttons on the far left to split the document for side-by-side or top-to-bottom comparisons; cursor option buttons for panning, selecting, zooming, and text selection; a dropdown menu for zoom level; page navigation buttons; and most importantly, the profile selection dropdown menu.

Figure 3-5: Navigation Bar

Toggling the Bars

Professionals frequently call the authors and other Bluebeam user experts in a panic because their menu bars have mysteriously disappeared. The truth is that the menu bars didn't mysteriously disappear, but were accidentally turned off by the user. Each menu bar is easily turned on or off with a single-function F key, and the user more often than not hit the mysterious key by accident. The keys are as follows:

- Command Bar: F9
- Status Bar: F8
- Navigation Bar: F4

Assuming the Command Bar has not been toggled off, each of the other bars can be toggled in the View-Interface menu, as shown in Figure 3-6. If the Command Bar has been toggled off, hit the F9 key and then the View-Interface menu will once again be accessible.

Figure 3-6: Menu Bar Toggle

Toolbars

As touched upon in chapter 2, toolbars are completely optional and customizable within Revu. Again, they typically line the perimeter of the view window and provide easy access to tools the user needs most. Toolbars will be discussed in detail in the "Document Navigation" section of this chapter.

Panels

Bluebeam developers realized having tools within one mouse click was advantageous to the user, but also acknowledged there were too many tools to keep the UI free from clutter without some form of organization. As such, Bluebeam utilizes a set of three panels, shown in Figure 3-7, that can be expanded or hidden from view by the click of a button. On the left, right, and bottom of the screen blue semicircles indicate a hiding panel. The user can single-click the blue semicircle to restore it to its previous expanded state, or click and drag the semicircle to expand the panel to the desired expanded state.

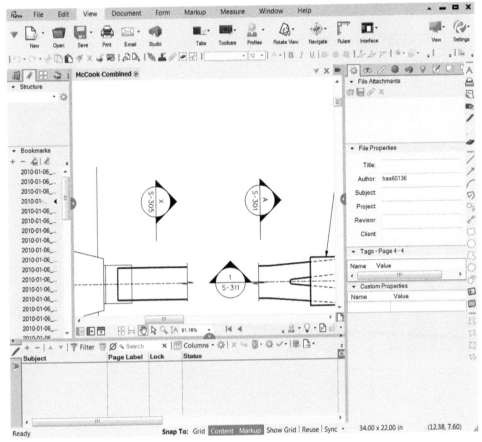

Figure 3-7: Panel Access

Tabs

The panels are the home of a series of tabs that store various tools, features, and UI functionality. Each of the tabs can be turned on or off, and as previously noted in chapter 2, any of the tabs can easily be dragged from one panel to another, allowing the tabs to be placed in the location most convenient for the user. As illustrated in Figure 3-8, to toggle tabs on, simply right-click in the empty space adjacent to the existing tabs, click Show, and click the appropriate icon to turn it on.

Tabs can also be toggled on in the View—Tabs menu, the tab access menu which is the orange down arrow in the upper left, or be turned off by right-clicking the tab itself and choosing Hide. The list below summarizes the themes and special features of each tab.

- File Access: Access recent or "pinned" files without having to navigate to them through a file/folder structure. Special features: Pin files to the recent list (see Expert Tip) and "Reopen files from last session" (see Figure 3-9), which restores the files open at the close of the last session.

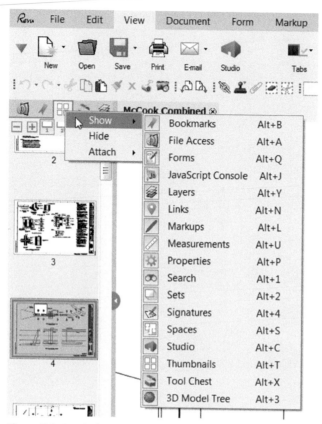

Figure 3-8: Panel Tab Toggle Menu

Figure 3-9: Reopen Files from Last Session Button

EXPERT TIP

Pin Files

Users of Bluebeam often like to reuse files that serve as "templates" for tasks they complete frequently. One specific example is a template file for a standard CD/DVD label. Bluebeam is an excellent tool for label making because it's incredibly easy to move and arrange logos, text, and other graphics. Unfortunately, a user may not create CDs frequently enough to keep the template file in the recent list but may struggle to remember where the template file is stored in the document management system. So he or she pins it. Simply right-click on the file in the recent list, select Pin, and click Pin File. This will move it to the Pinned Files section of the list and store the link there until the user specifically removes it. It's an exceptional way to locate files you come back to somewhat frequently. One can accomplish the same end by clicking the yellow pushpin icon that appears on the right side of the file in the recent list.

- Bookmarks: As covered in detail in chapter 2, add, edit, delete, or modify bookmarks in a file to help ease the navigation of large files. Special feature: Use "Create Bookmarks" to auto-generate bookmarks from content on the page such as text in the header, page numbers, or titles.

- Thumbnails: View, manipulate, and name pages of a file. The thumbnails tab is a very functional way to manipulate pages of a document. The tab can be used to reorder, rotate, delete, copy, add, replace, extract, number, or label the pages. Functioning similar to Microsoft's Power Point Slide Navigator, it's a very intuitive interface. Special feature: Use Create Page Labels to auto-generate page labels from content on the page. One very useful example of this is creating page labels from drawing numbers. Often page labels begin as a simple counting scheme (1, 2, 3 . . .). Unfortunately, when the user is looking at a set of engineering drawings, this type of page number is not highly valuable. "Page 3" generally doesn't give the user any valuable information about what is on the page. With the Create Page Labels feature, "Page 3" and all the other page numbers can be replaced with the respective drawing numbers such as "S-101" captured from content on the page. For the user, "S-101" indicates that the drawing is a structural drawing and that it is a plan view (series 100), both of which couldn't have been inferred from "Page 3." Additionally, the tool can be used to combine multiple page content entities into one page label. See Figures 3-10, 3-11, and 3-12 for an example of creating page labels with the tool. A video demonstration of the Create Page Labels process, or more generally, "AutoMark," can be found on Bluebeam's website at www.bluebeam.com/us/products/revu/12/automark-2.0.asp.

- Tool Chest: Store commonly used redlines in one easy-to-access location. The Tool Chest tab is one of the most powerful tabs in Bluebeam Revu. Tool sets allow users to save "tools" or redlines that they have created for reuse in the future. For example, a mechanical engineer may need to draw a pump symbol hundreds of times in a week. The tool chest allows that mechanical engineer to store the pump symbol for reuse. Users can create as many tool sets as they like, share those with a project team, or even network those tool sets such that changes automatically get distributed to the whole team. Additionally, users may

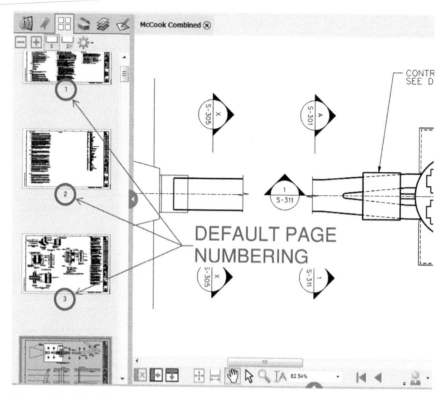

Figure 3-10: Default Page Numbering

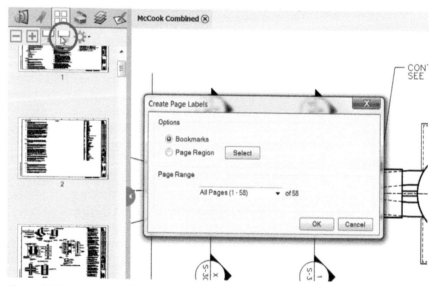

Figure 3-11: Create Page Label Window

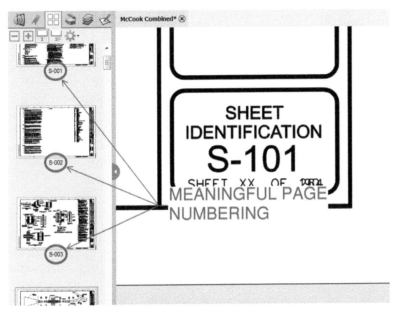

Figure 3-12: Meaningful Page Numbering

download a number of prepopulated tool sets from Bluebeam's website at http://support. bluebeam.com/revu-extensions/tool-sets/ or from the Tool Set Exchange Community Forum at http://communities.bluebeam.com/showthread.php?5038-Tool-Set-Exchange. Finally, the Recent Tools tool set will automatically capture the most recently used tools so they will be available to the user. Special feature: The My Tools tool set automatically creates keyboard shortcuts for each tool. When a user adds a tool to My Tools, a ghosted number appears above the tool icon. That number is the keyboard shortcut for the tool. If the user types the number on the keyboard, the tool will be ready to use with a simple click.

■ Layers: Create, delete, and control the visibility of any layers in the file. The layers can be used to separate various redlines. For instance, there may be separate layers for each engineering discipline: mechanical, structural, electrical, and so on. The markups or redlines for a given discipline would reside on the corresponding layer and could be shut off or turned on as needed. Special feature: Layers may be configured to turn on and off at specific zoom levels. So, as the user zooms into a drawing, layers can be configured to turn on as the user goes beyond the threshold.

■ Signatures: Locate, validate, and apply digital signatures within the PDF file. Digital signatures are the electronic version of "wet" signatures and utilize a digital encryption to verify and indicate that no changes have been made since the application of the signature. Digital signatures are explored in chapter 6.

■ Properties: Customize any individual markup in the file. Use the properties tab to change the color, line-weight, fill, text size and type, size, location, and so on. Every characteristic or property of a markup can be edited within the properties tab. Special feature: Though not tied to the properties tab, the Format Painter can be used to transfer applicable properties from one markup to another, simply select the source markup, click the Format Painter icon, and click the target markup.

- Search: As covered in chapter 2, electronically search through the file for a specific word, phrase, or even a graphic. In general, the tab functions as expected, though there are a few unique capabilities that don't exist in most search capabilities, the first of which is searching for a graphic. In Revu, the user can select a rectangular clip from the file and the program will scour the document looking for a matching pixel pattern. Special feature: In Revu, the user can perform a number of operations on the search results, both visual and standard text. Two examples are highlighting and hyperlinking, which means it is a very quick process to search and find all the pump icons, highlight them yellow, and add a hyperlink to the manufacturer's website.

- Measurements: Utilize various tools to scale, measure, and count. Combined with several features, the measuring and estimating tools are some of the most powerful capabilities in all of Revu. Chapter 7 is dedicated specifically to Revu's measuring, estimating, and take-off tools.

- 3D Model Tree: Navigate, edit, and control 3D PDF files within Revu. Chapter 10 takes a detailed look at 3D PDF files and how best to utilize them.

- Studio: Collaborate in real-time within your PDF files by utilizing Bluebeam's cloud platform Studio. Users may utilize the tab to create, invite, and control Studio sessions and projects. At the time Studio was released, it was groundbreaking and to date, nothing compares to the functionality offered. The details of Studio are covered in chapter 4.

- Links: Create and manage hyperlinks within the file. Though there are numerous methods to create hyperlinks within Revu, the tab greatly aids in keeping track of them. Special feature: Hyperlinks can be used to jump to many destinations including, external files, locations inside a file, websites, and specific Spaces within a file. In newer versions of Revu, those Spaces can be directly imported from an Autodesk Revit model, making setup very simple.

- Forms: Create and manage form fields within the file. Forms are discussed further in chapter 9.

- Sets: Create and manage Bluebeam Sets, a newer feature of Revu that is covered in detail in chapter 5. Like Studio, Sets completely changes the design flow by automating labor intensive and repetitive processes down to a few simple clicks.

- Markups: Summarize, status, and respond to redlines within a file. The markup list is at the heart of redlining within Revu and is both convenient and powerful. Chapter 4 takes a thorough look at the markup list, its functionality, and potential best practices.

- Java Script Console: Evaluate and control Java scripts within Revu. Though not specifically covered in this book, Java Scripts and their capabilities will be explored in the final chapter of the book where the authors share their speculations for the future of Revu.

Profiles

The beauty of the Revu UI lies in the Profile. One might compare the profile to the arrangement of a desk. Each professional has a preferred desk arrangement, which depends on the duties and tasks assigned to that professional. In one instance, the stapler may be on the left, sticky notes on the right, and pencil holder on the shelf. A neighboring colleague may prefer something totally different. The same is true for the layout of Revu; different users want different things. Luckily, the simplicity or complexity of the UI is completely and easily controlled by a customizable Profile. Though equipped with a selection of default Profiles, including Advanced, Simple, and Tablet, any changes the user makes to the UI are automatically saved to the profile and remain stored in the software for the future. The user can also create, name, and save a customized profile that can further be shared via e-mail or a network drive. Two examples are shown in Figures 3-1 and 3-2 at the beginning of this chapter.

The Profile sets everything from which toolbars are displayed along the perimeter, to which panel each tab belongs to, to which tool sets are available for use. It can change the look and feel of Bluebeam in a matter of two clicks. The power of this feature is that a user can jump from a simple task to a complex task and quickly adjust the UI to align with the task at hand.

The preloaded Profiles were created by Revu developers to arrange pertinent tools at the fingertips of a user who is completing a specific task. The Design Review profile arranges redline tools along the perimeter of the view. The Simple profile removes nearly every button in the UI. The Tablet profile optimizes the settings for a Windows tablet user.

The Profile can be selected using the down arrow next to the Profile icon (nicknamed Patrick) located in the Navigation Bar, as shown in Figure 3-13. The list will contain all the preloaded Revu profiles and any custom profiles created or imported into the software.

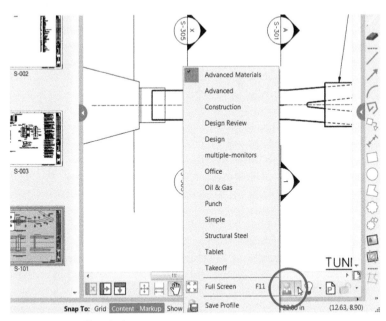

Figure 3-13: Profile Selection Menu

Document Navigation

The ease of navigating a document inside Revu is perhaps the difference that makes electronic markup a possible winner over traditional hard copy redlining. Versus the "clunky" navigation of competing software, nearly all users choose "flipping pages" as their preference. However, with Bluebeam, it becomes possible to move through a document with ease, and in many cases, faster than it's possible with paper. The following sections explore the ways Revu has made navigating simple.

Tabbed Navigation

Similar to the tabs on today's internet browsers, Revu utilizes tabbed navigation such that when more than one file is open, each file has a tab indicating its file name. The current or selected tab is the file that's displayed on the screen, and to switch between files the user only needs to click on the tab of the file of interest, as shown in Figure 3-14.

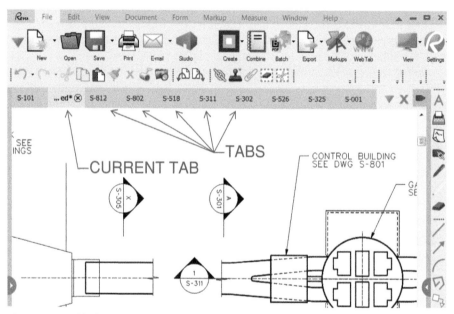

Figure 3-14: Tabbed Navigation Sample

Many files can be opened simultaneously, with each getting its own tab. As the files become numerous, the tabs become too small to be functional, and Revu begins to store the extra file tabs in a dropdown list on the right of the tabs. The list is marked by a down arrow, or downward pointing triangle, and may be accessed simply by clicking that arrow, as shown in Figure 3-15.

The tabs also contain a little bit of file functionality. In all cases, clicking the X button on the right end of the tab will close the file. In cases where the file resides in a document management system such as Bentley ProjectWise or Microsoft SharePoint, a symbol will appear at the left end of the file tab. That symbol indicates the state of the document, whether it's checked out, locked, or the like. Clicking the symbol accesses a dropdown menu that gives the user options on what to do with that document, as shown in Figure 3-16.

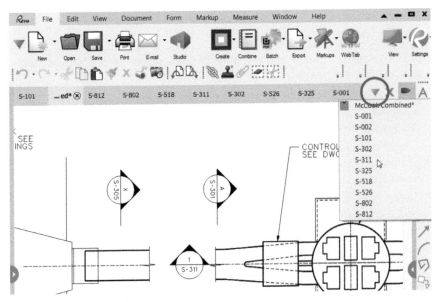

Figure 3-15: Tab Access Menu

Figure 3-16: File Tab Document Management Features

Tricks with Tabs

While tabbed navigation is pretty helpful by itself, there are several related features that also improve the user experience.

1. As the user edits a file, an asterisk appears next to the file name indicating that the file has unsaved changes. When the file is saved, the asterisk disappears. As such, the user has a quick reference to know whether the file needs to be saved or not.

2. After a few hours of work, avid users may find themselves with many files open even though they are done with all but the file currently being edited. A quick way to purge the unneeded files is to right-click on the one tab currently in use and select Close Other Tabs. This will close all the files except the selected one.

3. Sometimes the user wishes to close the open files, but leave Bluebeam open. Under the Window menu, there is an option called Close All, which will close all of the open files, but leave the instance of Revu up and running. Note: This is especially helpful in cases where the user incidentally opens tens or hundreds of files and clicking the close button on each file tab would be excessively time consuming.

The Mouse

In Revu, the mouse combined with the nondominant hand on the keyboard can facilitate nearly every navigation feature without traveling through menu after menu to find a button that unlocks one single feature. The mouse goes further than navigation though; it also enables many capabilities and features within the various markup tools. Simply put, the user can do a lot without ever clicking a menu. Some of the mouse navigation features are listed below.

■ Scroll and Zoom: From chapter 2, readers know that Revu has two viewing modes, One Full Page and Scrolling Pages. When the user opens a document-style file, Bluebeam defaults to Scrolling Pages mode assuming the user wants to scroll through the document. When the user opens a drawing-style file, Bluebeam defaults to One Full Page mode assuming the user wants to zoom in to look close at a detail. Out-of-the-box, the mouse wheel scrolls in Scrolling Pages mode and zooms in One Full Page mode; however, those default settings can be customized in the preferences window (Ctrl + K), as previously discussed. Zooming can also be accomplished by changing mouse to the zoom cursor, one of the four buttons found at the bottom center of the UI in the Navigation Bar, as shown in Figure 3-17.

■ Pan: The mouse can be used for panning in a document when the pan cursor is selected (Figure 3-17). The user simply clicks and drags the page to move to a different part of the page.

■ Select: Selecting markups or redlines can be done using the pan cursor by clicking an individual mark. Similar to other software, multiple markups may be selected simultaneously by holding the Shift key at the same time, or the select cursor, also shown in Figure 3-17, may be used with a left click to draw a rectangle around or through the desired marks. Swiping from left to right includes only those marks that are entirely enclosed by the

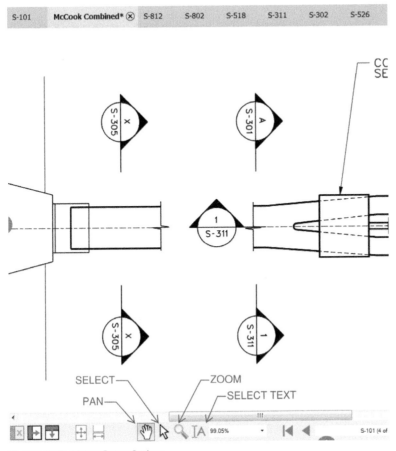

Figure 3-17: Mouse Cursor Options

selection rectangle, and swiping from right to left includes any marks that are enclosed or intersected by the selection rectangle.

- Select Text: The final mouse cursor option is the text selection tool that can be used to highlight or select a segment of text within a document for copying, highlighting, and so on.

EXPERT TIP

Empower the Mouse

As much as the mouse can do by default, it can do even more with some help from the keyboard, the right mouse button, and some custom settings. First, the authors have found that users prefer the default mouse wheel behavior to be consistent regardless of page mode. As such, we recommend choosing scroll or zoom as the default for both page modes based on the user's individual preferred behavior. Second, the behavior of the mouse wheel can be switched by holding the Ctrl key on the keyboard. So, if the default is scrolling, the Ctrl key will cause the wheel to zoom and vice versa. Third, right-clicking the mouse with the pan cursor enables the select cursor without having to click the cursor icon. Fourth, the Shift key may be held to cause the scrolling behavior to move the document left and right instead of up and down.

Throughout Bluebeam, the mouse can be empowered by utilizing secret keys on the keyboard. This book will point out many of them, but others will have to be discovered by the reader. As a guide, the reader may find a list of all keyboard shortcuts in the Help menu.

Splitting the Screen

As noted in the introduction to this segment, traditional redlining consists of "flipping pages." Professionals who are reviewing drawings or other documents like to flip from one page to another, and in some instances these two pages of interest are many pages apart. In engineering work, the plan drawings often appear near the front of the set and refer to detail drawings that appear near the back. Working in hard copy form, it's easy for the reviewer to hold both pages and quickly go back and forth. Working in traditional electronic review software, it's significantly more time consuming as the user either needs to scroll through the middle pages or remember the two page numbers and type them in each time. In either method, the routine process of going back and forth is more difficult.

Revu put an end to that by enabling the ability to split the screen. By selecting the Split Vertical or Split Horizontal buttons in the bottom left, the current document will be duplicated in two windows, allowing the user to be in two places within the same document at the same time. Now the user who wants to flip between a plan sheet and a detail sheet only needs to split the screen and move his or her eyes back and forth between the two windows, as shown in Figure 3-18.

It's important to note that the windows can be split multiple times, each time dividing the current window into two equal parts, either horizontally or vertically. Suddenly "flipping pages" is no longer a hurdle because it's easier for the user to move his or her eyes than physically change the page.

Flyout File Window

Though splitting the screen takes the user a long way in efficiency of redlining, users will quickly notice the shrinking screen real estate. With every split, the drawing file window gets smaller and smaller until eventually the window is too small to be functional.

The Bluebeam developers responded to that problem with the Flyout File Window, or Breakaway Pane. After splitting the screen, a user can left-click on a file tab and drag that file window out of the normal Bluebeam windowpane, as shown in Figure 3-19. This is exceptionally valuable for users with multiple screens because it allows them to maximize the use of their screen real estate. The same can thing can be done when more than one file is open; simply left-click on the file tab desired and drag it out of the Bluebeam pane. The user will also note that new tabs may be added to existing flyout windows by following the same process.

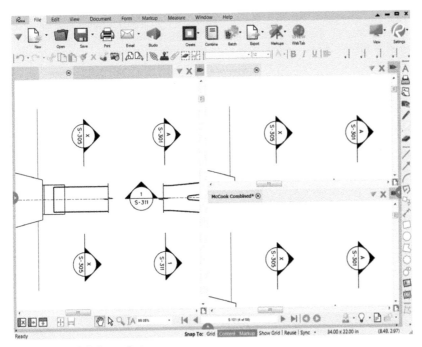

Figure 3-18: Split Screen Feature

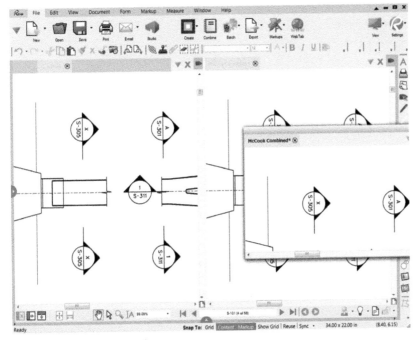

Figure 3-19: Revu Flyout Window

To remove the extra flyout window(s), either click the close button on all the tabs in the window or click and drag all the flyout tabs back to the main window. In either case, the extra window will simply disappear with the last file tab. Note that the user cannot move a flyout tab to a window that already has the same file open.

Sync Them Up

One of the most widely renowned features of Revu is Sync mode. In Sync mode, the documents in two distinct windows are synchronized so they scroll and zoom together. This is especially helpful when comparing two versions of the same file. As the user moves the file in one window, the second window mirrors that movement so it's easy to pick out any changes, verify updates, and backcheck redlines.

The Sync mode can be turned on by clicking Sync in the Status Bar at the bottom of the screen. The Sync will be highlighted blue, indicating it's turned on, and the two windows that are open will begin to mirror each other. To turn Sync off, simply click the Sync button again and the background of the button changes back to black.

Sync also has two modes, Document and Page, which can be selected using the down arrow on the right end of the box. In Document mode, the two files move through pages in synchronization, such that if file 1 is on page 10, file 2 is also on page 10. However, there are cases when a user would like to sync files with differing quantities of pages. That's where the Page mode becomes beneficial. In Page mode, the individual pages are synchronized for zoom and scrolling, but as one document scrolls to the next page, the second document does not.

At this point the user can utilize multiple screens with multiple pages and multiple files, making a compelling case for electronic redlining and markup, but there are even more tools at the user's disposal. When done correctly, thumbnails, page labels, and bookmarks are great assets in navigating a large file. All three can be accessed in the panels noted previously. Searching is another great feature of electronic markup. When a user needs to mark every instance of a given word or symbol, he or she can very quickly identify each location where it exists. Of course there are even more, some of which will be highlighted in later chapters of this book.

Toolbars

No doubt, Revu is loaded with toolbars. In fact sometimes, new users feel overwhelmed by the sheer number of buttons available to them in the Bluebeam window. As users grow comfortable with Bluebeam though, they will recognize that the buttons are there for easy access. There is no need to drill through multiple menus when the options are available with one click.

Though there are toolbars serving many functions in Revu, this section will focus on commands in the alignment and markup categories since those are most pertinent to redlining.

Out-of-the-box, most of these commands are housed in the Alignment, Shapes, Sketch, and Text toolbars, but can be added to any of the toolbars.

Depending on the profile a user is running, the mentioned toolbars may be turned on or off. Changing the profile is one way to change which toolbars appear, but it's also very simple to turn any of them on or off from any profile. As shown in Figure 3-20, a simple right-click in the margin space along the top or right side will bring up a list of available toolbars with check-marked orange boxes indicating which ones are currently on. Simply click the box to display them or remove them. The user will also notice the Customize option at the bottom of the menu. It is here where individual commands can be added or removed from a toolbar and also where custom toolbars can be created. The Customize menu is shown in Figure 3-21. At the top of the left side of the menu, categories can be selected from the dropdown box. As noted previously, the bulk of the redlining commands are in the Alignment and Markup categories. By selecting a category, the user can see a list of commands that can be added to a menu selected in the top right pane. The bottom right pane shows the commands currently displayed in that menu. In addition to the commands included in the menu, recall from

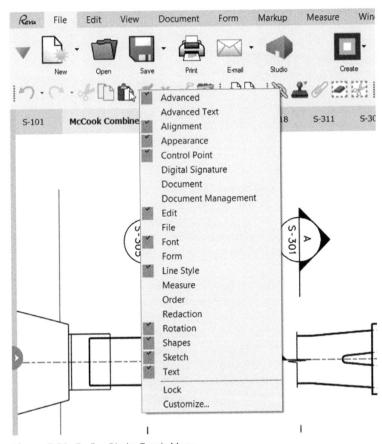

Figure 3-20: Toolbar Display Toggle Menu

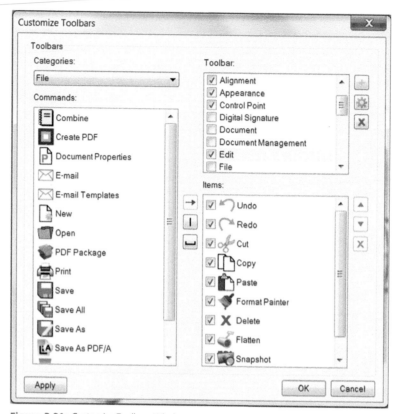

Figure 3-21: Customize Toolbars Window

chapter 2 that the location of the menu may be modified by left-clicking on the (. . . .) symbol and dragging the toolbar to the desired location in the margin.

Once the toolbars have been customized and located effectively, the redlining can begin. Because the markup commands are too numerous, this book will not address them all in detail. It will, however, highlight some easy ways to access the commands and share several tricks for using them well.

First, each of the commands has been assigned a keyboard shortcut. As noted in an expert tip, the keyboard can magnify efficiency because it can prevent the user from the back-and-forth motion of accessing a toolbar command with the mouse. To help out the user in learning the shortcuts, Bluebeam has identified each shortcut in parenthesis when hovering over the command. Most shortcuts are intuitive, such as T for Text Box and H for Highlight, but others are a little more creative. The letter Q, for instance, initiates the Callout command since C was already used for Cloud. Although it seems strange at first, the letter Q actually resembles a callout, which is a text box with a leader on it. The Polyline, shortcut Shift+N command is similar. Polyline does not start with N but the letter N easily looks like a Polyline. As such, the keyboard shortcuts are fairly easy to remember and can be extremely valuable.

Second, not only does the keyboard provide easy access to commands, it can be used to unlock hidden features of those commands. The Line command (keyboard shortcut L) as one example can be constrained to a vertical, horizontal, or 45-degree angle simply by holding the Shift key. The Rectangle command (keyboard shortcut R) can be constrained to a square also by holding the Shift key. The Ctrl key can be used with the Highlight tool make the tool function like a pen tool even when highlighting over text that it normally wants to select. The secrets are numerous and the authors encourage users to experiment. Hold the Shift or Ctrl key and see what happens. A short list has been provided below:

Shift

1. Add or remove leaders on a callout

2. Constrain a rectangle to a square

3. Constrain an ellipse to a circle

4. Move a markup directly vertically or horizontally

5. Add control points to a cloud, polyline, polygon or rectangle

6. Rotate markups at finite degrees

Ctrl

1. Click and drag a copy of a selected markup

2. Toggle the zoom/scroll mouse wheel setting

3. Group selected markups with Ctrl + G

Alt

1. Move an entire callout at once, not just the arrow or text box

Finally, each markup produced is easily customizable. From color to line type to size, Revu makes it much easier than most other electronic markup tools to change the format and look of a mark. The easiest way to modify the appearance is in the Properties tab, discussed at the beginning of this chapter. That tab can be accessed from the panels or opened by right-clicking on a given markup and selecting Properties. There the user will find all the formatting and appearance options for the given command. A sample Properties menu is shown in Figure 3-22 for the Cloud+ tool, a combination of a Cloud and a Callout.

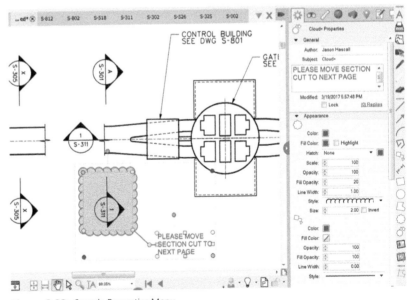

Figure 3-22: Sample Properties Menu

Save a Style

Just as people have certain tastes in furniture, individual Revu users have certain preferences in the way their individual markups look and feel. Some users like large font with boxes around every line of text. Other users like to fill the background of text boxes so the text is easier to read. Users who find themselves repeatedly changing the properties of a markup should consider setting that style as the default.

There are two ways to do this in Revu. The first is very intentional and apparent: simply select a markup, change the properties to those desired, scroll to the bottom of the Properties tab, and click the Set as Default button. The second is less apparent and often happens by accident. Simply initiate the command by clicking or keyboard shortcut, change the command properties in the Properties tab, and then place the markup. The next time the user initiates that command, the new style will be retained. Changing the properties after the markup has been placed does not retain the style as default.

Tool Chest

The final section of this chapter focuses on the tool chest. As noted in the initial discussion on tabs, the Tool Chest enables the ability to store, reuse, and share common markups.

A number of tool sets are included out-of-the-box and can be turned on with the properties menu at the top of the Tool Chest tab. Clicking the dropdown arrow allows quick toggling and clicking the icon itself launches a pop-up window where the user can import, export, remove, move, or create a tool set. The power of the tool chest is that complex markups like pump icons and engineering seals no longer need to be created or copied from a separate file each time they are used.

By default, each user is given two user specific tool sets, My Tools and Recent Tools. Recent Tools keeps a running record of all the markups and tools recently used such that the user could return to them easily if needed. The number of commands stored is customizable from 12 to 30. My Tools, on the other hand, is a dedicated place where the user can store frequently used tools.

Creating a tool is as easy as creating a markup, and those markups can be as complex as the user desires because multiple markups can simply be grouped. The engineering seal shown in Figure 3-23, for example, was built entirely from markup commands in Bluebeam and has more than 50 individual markups all grouped together. (Note: The authors do not recommend building seals in Revu, but have done this for an example of a complex markup.) Once a markup or group of markups is ready to become a tool, the user simply right-clicks on the markup or group and selects Add to Tool Chest, and then selects the desired tool set from the pop-up list. The possibilities are truly endless.

One of the more recent and advanced features of tool sets is the ability to assign a scale. Doing so causes the tools in the tool set to automatically resize based on the scale of the document or viewport. So, in the example shown in Figure 3-24, a wide flange tool drawn to scale automatically resizes itself for use in markup or sketching the particular connection detail when it's used inside of a scaled region called a Viewport. Outside of that region, the

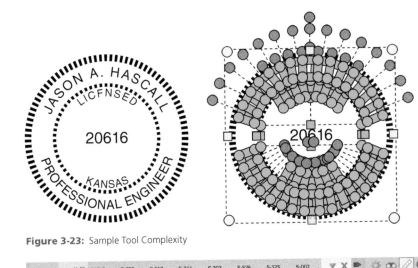

Figure 3-23: Sample Tool Complexity

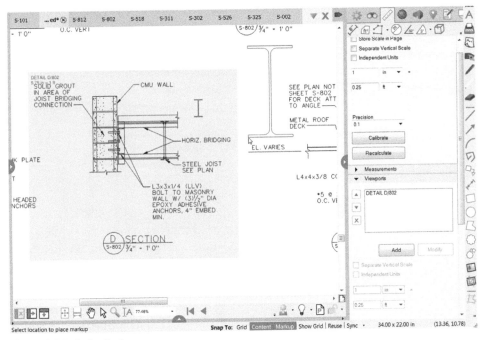

Figure 3-24: Scaling Tools

markup is scaled to the default page scale. A given tool set can be calibrated by choosing the Set Scale option found in the individual tool set properties menu. A tool can be drawn to scale utilizing the Sketch to Scale commands.

Finally, the newest feature of tool sets is the ability to create a legend of the tools in a particular tool set. Also under the tool set properties menu is the Legend button. Clicking the button displays the option to Create New Legend, which adds a legend of tool set mark-ups to the page. The Legend is entirely customizable and allows the user to keep track of the quantity of each markup on the page. For instance, as shown in Figure 3-25, if a floor plan was being created, the Legend could keep track of the total number of square feet of each type of slab or flooring. Legends can also be used to gather information from marks belonging to more than one tool set.

Don't Reinvent the Wheel

Many times tools a user would like exist in CAD or BIM drawings but aren't easily re-created with the markup capabilities of Revu. In these cases, Revu's Snapshot (keyboard shortcut G) can become an incredibly helpful tool. Unlike many screen capture tools, Snapshot captures vector data when it's available. So CAD or BIM drawings that are typically vector can be captured into a custom tool simply by taking a Snapshot. Once taken, the Snapshot can be pasted onto the page and the colors of the vector content may be edited to stand out.

For example, if a piping and instrumentation diagram has three types of pump symbols indicated, each symbol could be captured with Snapshot and pasted onto the document. Once pasted, Change Colors can be used from the Properties tab to convert black linework into red, blue, or another color, causing the pump symbols to stand out on the page. The images can also be Cropped as needed. From there, each of the three symbols can be added to a new or existing tool set as discussed above.

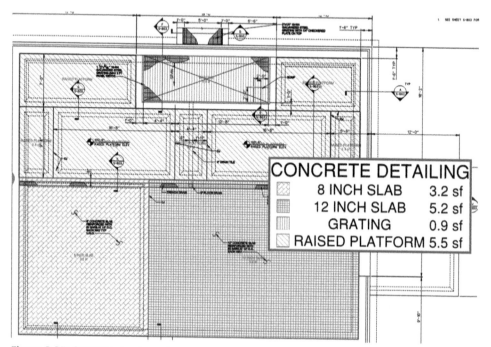

Figure 3-25: Sample Markup Legend

Meet Brett Agee, Project Engineer, Bluescope Conventional Steel Services

When Revu 2016 was released, Brett Agee saw one new feature that he thought could change his world, Legends. He had the perfect application and he knew it would make his life easier.

Brett is a structural engineer who works for Bluescope Conventional Steel Services, a steel building manufacturer in Kansas City, Missouri. Bluescope specializes in designing, fabricating, coordinating, and shipping a complete building package to a job site, where it is then constructed by one of Bluescope's thousands of approved erectors. Brett's role is specifically on the design side and he often has to interface with clients, architects, engineers, suppliers, and fabricators.

After watching the 2016 launch demonstration, he was confirming an estimate from a deck manufacturer. This particular project had numerous deck openings of various shapes and

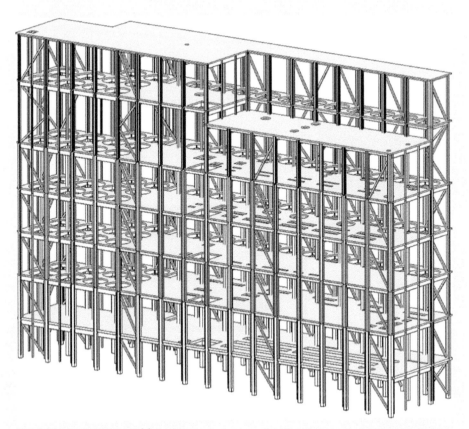

Sample Bluescope Project Model

sizes. Since each opening required a pour stop angle around its perimeter, that particular line item was more substantial than his typical project. Brett suspected the supplier had overcompensated to account for tolerances, unknowns, and construction errors. Using the new Legends tool was his first thought. He knew he could easily check the supplier's quantities, and that's exactly what he did. "I decided I could do it the way I've done it before, or I could save time and do it this new way," Brett said.

His idea was to make a visual map of all the pour stops on the drawings and then use a legend to calculate the totals. "I thought it would be a really nice way to create and update the total length of each type of pour stop," he said. In practice, the idea was pretty simple. Choose a color for each type of pour stop, trace each of the openings with a perimeter measurement tool, and sum the totals with a legend. As revisions occur, he would simply be able to copy the perimeter elements to the new revision and update the tracing as needed. The legend would take care of the totals automatically.

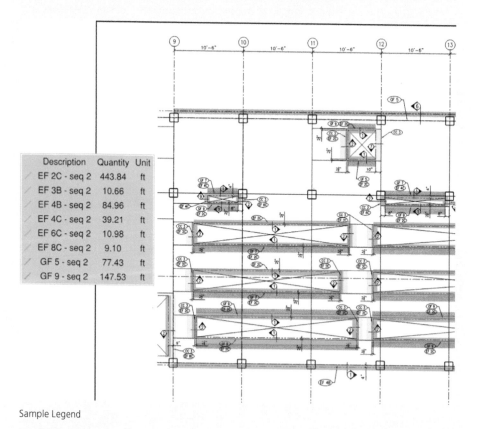

Sample Legend

As it turned out, Brett was able to save about 30 percent of the pour stop cost on the project and lower Bluescope's estimate for the proposal. He also approximated the time savings for completing the estimate at 50 percent compared to the old way of estimating with spreadsheets.

Today, Brett uses the legend tool for all sorts of quantities. From paneling to roof deck to joists to openings, if it's measured in square footage or length, he uses a legend. But he's not the only one at Bluescope using legends. Brett noted, "Several of the quoting engineers have started using it, probably about 5 of them, but it's growing. When they see it, they immediately want to use it. After a 5-minute demonstration, they're off and running."

Brett finds the use of legends easier than spreadsheets in almost every way. Visually it's better. He can see every measurement directly on the drawing, not only to verify the accuracy of the measurement but also to verify that no elements have been missed or counted twice. Thanks to Bluebeam, it's also easier to set up. Once a tool set exists, creating a legend is only a few clicks, so he creates a tool set for each group of items he measures. Finally, the legend updates automatically and instantaneously, something that wasn't true with spreadsheets.

While the use of legends is growing and growing, Brett and the rest of the Bluescope team are also looking for other ways to use Revu. "The Revu 2017 Quantity Link is pretty exciting to us," said Brett. "We use Microsoft Excel for many things, so my colleagues and I are happy about the introduction of that feature."

One example where they are putting the new Live Link to use is with standard connections. Bluescope has a library of typical connections and each one has an identification number. A number like 432A might correspond to a specific type of shear tab. Each of those standard connections has corresponding standard quantities: quantity of bolts, area of plate steel, or length of weld, for example.

The connection number is printed on the Bluescope drawings everywhere that connection occurs and Revu can be used to search autonomously for those connection numbers and count them. Thanks to the new Live Link, that count can be tied to Excel and used to calculate the subordinate quantities associated with the connection type.

"We have the Excel link working, but it's not quite ready to roll out yet." Brett noted. "Our design software is supposed to count all the bolts, welds, and plates, but this is a good, easy way to double-check and also a way to check historical sets where electronic files don't exist."

Not surprisingly, the Bluebeam impact hasn't stopped at the door. Revu has helped Bluescope collaborate with their clients, partners, and vendors.

A typical project for Brett begins with architectural drawings. To get some quick area quantities for vendors, he can use the scaled drawings and mark the areas directly where they should be. Without having to build a spreadsheet, the quantities can be shared with the vendors both visually on the drawing and with total quantities in a legend. The color-coded marks have also helped with quality assurance because they're so visually indicative that on a few occasions they have tipped off the architect that Bluescope has misunderstood the design. For example, brick facade may have been mistaken for stucco

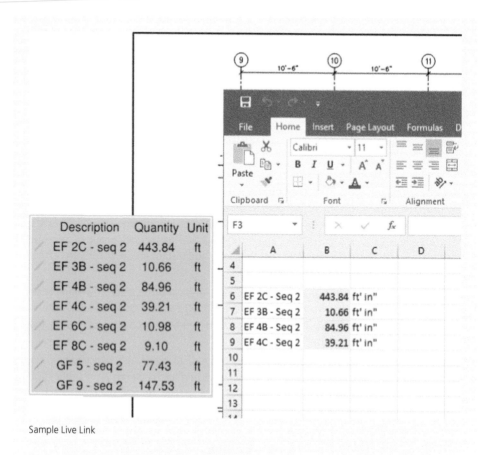

Sample Live Link

and blue panel might have been replaced by gray panel. Those types of mistakes are minimized because of the visual nature of the markups.

It also helps with scope discrepancies. Maybe there are two wall vendors, each responsible for a different type of panel. When all the panels are color-coded according to vendor in the PDF, it's easy to know which panels fall in which vendor's scope. It also creates good bookkeeping such that no panel is omitted and no panel is counted twice.

Brett also has good luck using Revu on conference calls with screen shares. He noted that multiple parties are usually represented on the calls and each of them is using a different jargon or terminology for items or locations in the project. Visual color-coding is very easy to do and quickly cuts through all the jargon, resulting in very clear communication. "If I highlight a brace red in multiple views, it's easy to know and remember where the brace is in each view. People find it easier than trying to follow along when it's not highlighted. We were recently working with a theater consultant. She didn't understand all the terminology we were using. It was inconsistent. The color-coding really helped on that project," Brett said.

Bluescope's deck manufacturers appreciate the PDF markups and legends too. He noted that they've been receiving PDF files that have been redlined really well from many of their partners. They've become so accustomed to the new format that it has significantly reduced the number of questions; they just understand it and trust it.

As noted previously, manufacturers and suppliers usually include a safety factor to cover themselves and make sure they aren't under on quantities. When safety factors from multiple vendors, suppliers, and Bluescope compound with each other, those safety factors can become inflated. When Bluescope provides the exact quantities for the estimate, they can have full control of the overall safety factor for the project, helping them be more competitive on their proposals.

Finally, Brett believes that Bluebeam still has room to grow at Bluescope and he's excited to see where it goes. He also believes there's a critical mass usage of Revu within his company that they haven't quite hit. "Once we get over that line, the sky is the limit. Our users just need to think of what else they can do."

Conclusion

Although Bluebeam Revu is improving each year, introducing revolutionary features that automate processes, simplify tasks, and improve value of deliverables, redlining was the headlining capability and, in the authors' opinions, is still the greatest single capability of Revu. As this book progresses through the higher-order capabilities of the software, readers will identify how even the most advanced features tie back to redlining in some way.

Chapter 4
Redlining Together

One of Bluebeam's best-kept secrets is Studio. This collaboration suite is modestly hidden under one small button (Figure 4-1).

Amidst Bluebeam's hundreds of features, it can be easy to miss this powerful tool. Defining the value proposition for Studio sounds a lot like the major theme of this book—performing work functions better with digital. Think about how a team marks up a set of documents in the physical world. They probably have the master set of paper documents sitting on a table in a common area or room. When someone needs to record a mark, he or she gets up from the desk, walks over or up or down or wherever to the master document, hand-draws the correction, and documents a few pieces of information about the change in a paper log. When Engineer Jane is working on the document, no one else can touch it. When Engineer Joe forgets to log the change information, rogue marks appear on the document with little leading back to their origin. When Engineer Javid in Pune needs to markup the document, he has to request a copy of the master set, print it out, mark it, scan it back in, and send it back to be copied onto the master set. The traditional methods of collaborative document markup lead to higher potential for errors, unnecessary time delays, and inefficient use of physical materials.

Enter Studio.

Studio allows project teams to collaborate on documents anytime from anywhere in the world. Think of it as a virtual room. Instead of having to physically transport oneself to the documents (or fake it, if one is not in the same office building), a collaborator can open up the one source of truth document on his or her own device and edit right from his or her location. Changes made to the documents in Studio are always made to the master documents, because there is only one set of documents. To fully explain the capabilities of Studio, it's important to start with the distinction between Studio Sessions and Studio Projects.

A Studio Session is made to collaborate live on PDFs. In a Session, as many as 500 collaborators can be marking up the same document simultaneously, from anywhere in the world. Each markup is automatically logged with data such as author, creation date, page number, and more in the ongoing Markup Log for each document in the bottom panel. A Session can hold up to 5000 documents at a file size of 1 GB each, and a markup itself can even have a size of up to 10 MB. With Sessions, there is only one version of each PDF document; no version history exists, besides the record of each individual markup.

A Studio Project is a light document management system with some similar capabilities to the basic functionality of SharePoint, Documentum, ProjectWise, and Box: a check-out/check-in system with version history. In a Project, files of any type can be uploaded and

Figure 4-1: Studio Button

shared, not only PDFs as is true with Sessions. There are no limits on the number of collaborators, number of files, size of files, or size of markups in Projects. In a Project, collaborators can only work on files one at a time. Sessions can be launched from Projects so that PDFs in a Project can be worked on live for a period of time and then saved back into the Project as a new version, complete with all the markups from the Session.

The major differences between Sessions and Projects are shown in Figure 4-2.

Studio Session and Projects Comparison

	Sessions	Projects
How many attendees can join?	500	No limits
How many files can we upload?	5,000	No limits
What files does it support?	PDF	Any file format
How big can the files be?	1 GB each	No limits
How big can the markups be?	10 MB each	No limits
How much space can we use?	No limits	No limits
Can attendee access rights be managed?	Yes	Yes
Can previous file revisions be viewed and restored?	No	Yes
Can we receive notifications about file and user activity?	Yes	Yes
Can we send and receive markup alerts?	Yes	No
Do all attendees need a license of Revu?	No	No

Figure 4-2: Differences Between Sessions and Projects

A person considering using Studio should think about the functionality required for document collaboration to decide if Sessions, Projects, or a combination of both would be best for the specific use case. Some things to consider:

- Is the project team far-flung? If so, Studio Sessions is likely the most efficient means of collaboration on PDFs, beating geographical and time zone limitations.
- Is the project team made of different entities? For team members from different companies, Studio Projects may help to provide the team with a common working document repository, which requires only a Bluebeam Studio account to access and avoids the entanglements of adding outside personnel to any one company's IT network, or negotiating whose existing system is the best to use.
- Does the project have tight deadlines on review cycles? If so, Studio Sessions can save weeks in turnaround time by having reviewers simultaneously work on PDFs.
- How tech-savvy and willing to learn is the project team? Finding a great tool like Studio is only half, or maybe a third, of the challenge of improving project workflows. A large chunk of effort must be spent on training, change management, and setup when switching from a paper process to digital.

Getting into Studio

Once a decision is made to use Studio in any capacity, the first step is to create a Studio account for each user. The Studio account is based on the user's email address; a person could conceivably have as many Studio accounts as he or she has email addresses.

Two other Studio options to note here: First, Studio Server by default is studio.bluebeam. com. This will be the only option unless a company decides to purchase Studio Enterprise. Studio Enterprise lets a company use its own servers to host the Sessions and Projects, storing files securely behind the company's own firewall. This would be a good option for companies who comply with NERC-CIP or FedRAMP data security requirements. There are several other benefits to Studio Enterprise, which can be seen on Bluebeam's website www.bluebeam. com/us/products/studio-enterprise/. The last checkbox on the login dialog box also relates to Studio Enterprise. If a company runs Studio Enterprise on its own IT system connected to its own Active Directory, its users can check the Use Windows Authentication box to auto-matically populate their passwords from their company Windows account.

Second, Studio Prime is a higher level of Studio administration that allows a company more control over its Studio users. Prime enables user management and visibility into data analyt-ics for the organization's use of Studio. Also with Prime, an organization can write custom scripts to automate repetitive functions for files within Studio, like moving files, converting files to PDF, adding cover pages, and more. For more information, see Bluebeam's Studio Prime webpage: https://www.bluebeam.com/solutions/studio-prime.

To the right of the Start button is the Join button. This button allows the user to join another Session or Project by entering its nine-digit ID number. To the right of the Join button is the Settings button. This button opens the Studio section of the Preferences dialog box. Here, the user can manage Notification Preferences, request a password reset, manage the available Studio servers, logout of Studio, and edit some general Studio options (Figure 4-3).

Figure 4-3: Studio Preferences

EXPERT TIP

Studio Preferences

The Studio section of the Preference dialog box shows four General preference checkboxes:

- Checkout on Open
- Enable Flashing on Session Alerts
- Toolbar Integration Only
- Force Proxy Use

"Checkout on Open" applies only to Studio Projects, not Sessions. If this box is checked, it means that when a user opens a file from a Project, the file will automatically be checked out in the user's name. This can be helpful if the user is more likely to edit a file upon opening it than simply viewing it. However, it can be annoying if the user opens files frequently for viewing and then has to cancel the checkout for each file instead of just closing it.

"Enable Flashing on Session Alerts" applies only to Studio Sessions, not Projects. If this box is checked, it means that when a user is in a Session, every time another user performs an action, the Session toolbar will flash orange, kind of like when you receive an Instant Message through Lync or Skype for Business. With the possibility of up to 500 collaborators working simultaneously in a Session, this is a helpful box to know about when working in Sessions.

"Toolbar Integration Only" applies to opening and saving PDFs, both inside and outside Studio. If this box is checked, it means that when a user clicks the Open button on the main toolbar, the dialog box will default to opening a file from the user's computer.

Open

The same is true when saving a document; clicking Save or Save As will automatically open the user's computer file structure to save the file to the local hard drive. Conversely, leaving this box unchecked means that the user will be given the option of opening the file from (or saving the file to) any integrated document management system. Currently, this means the user's hard drive, any connected ProjectWise directories, any connected SharePoint libraries, or any existing Studio Projects.

If a user is working frequently out of one of the mentioned document management systems, it may be worth leaving this box unchecked so the user can easily open from and save directly to these systems without having to save the file to his or her hard drive first and then uploading it to the correct system.

"Force Proxy Use" relates to how Studio connects to a user's computer through the company's server. If the server is configured correctly, Studio will automatically connect through that proxy. If Studio is not automatically connecting, the user can try checking this box to force connection through the company's proxy. If connection still is not successful, the user may have to ask the company's IT department to unblock the Studio hosts. Check Bluebeam's Support article on this issue for more information: http://support.bluebeam.com/articles/error-could-not-connect-to-bluebeam-studio/

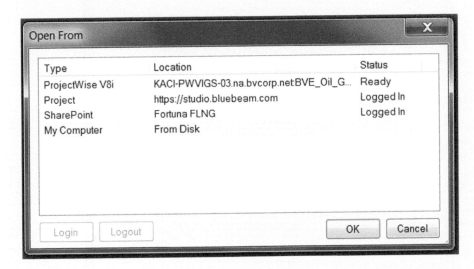

This method of joining a Session or Project will only work if:

1. The owner of the Session/Project already allowed the user's email address to enter, or

2. The owner of the Session/Project did not restrict access by email address.

There are two prompts that would cause a user to join a Session or Project. First, the user wants to create one him- or herself, or second, the user has been invited to one by another user. Once a user has joined a Session/Project by either means, it will appear on the Studio home page and can be rejoined by clicking on the Session/Project name. At this time, a user can only work in one Session or one Project at a time in one instance of Bluebeam. It is technically possible to work in multiple Sessions/Projects at once by opening multiple instances of Bluebeam on the same device. However, a user cannot use the same Studio account to work on multiple devices at once. Logging in on one device will log the user out of the first device.

Figure 4-4: Start Session or Project

To create a Session or Project, click on the Start button on the Studio home page. Choose either a New Session or a New Project (Figure 4-4).

Sessions

Starting with Sessions, the Start Studio Session dialog box will appear (Figure 4-5).

Type in a name for the Session in the Session Name field. Select documents to add to the Session by clicking on either the Add Open Files button or the Add button below the Documents pane. Add Open Files will load all currently open PDFs into the Documents pane.

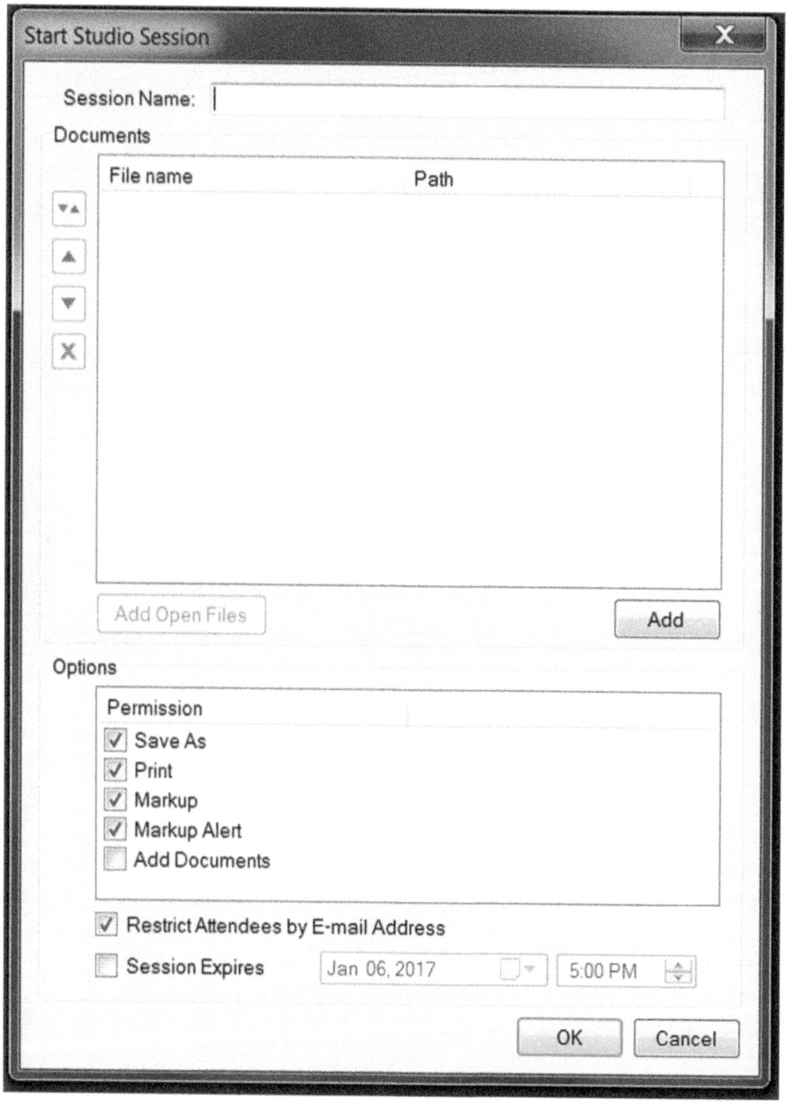

Figure 4-5: Start Studio Session Window

Add will allow the user to select PDFs from his or her hard drive, if Toolbar Integration Only is checked in the Studio section of the Preferences dialog box. Add will allow the user to select PDFs from his or her hard drive and any other integrated document management systems, if Toolbar Integration Only is unchecked. The user can use a combination of Add Open Files and Add to get all the desired PDFs ready for upload.

Once the documents are arranged, consider the default Session permissions, which are listed in the Options pane. These permissions will apply to everyone who enters the Session, until the Session owner modifies permissions inside the Session settings. These permissions should be set at the level of the most restricted collaborator, since they dictate what a collaborator will be able to do upon first entering the Session. The owner of the Session can set up more capabilities for other groups of users after creating the Session via the Session settings.

- Save As: Allow the collaborators to save a copy of the documents to their local hard drives, or to other integrated document management systems.

- Print: Allow the collaborators to print a copy of the documents to paper or to PDF.

- Markup: Allow the collaborators to mark up the documents.

- Markup Alert: Allow the collaborators to send email notifications about markups to other collaborators.

- Add Documents: Allow the collaborators to upload additional PDFs to the Session.

At the bottom of the Start Studio Session dialog box, there are two more options. The first, Restrict Attendees by E-mail Address, limits Session access to those whom the owner invites by email address. If this box is unchecked, any user with the nine-digit Studio ID can join the Session via the Join button. The Session creator can still restrict certain users from entering, or explicitly allow others to join, by adding users to the Attendee Access list and selecting the Allow and Deny options. It is recommended in most situations to keep this box checked. The second, Session Expires, allows the Session creator to set an expiration date by date and even time. When a Session reaches its expiration date, it will close to every collaborator except the owner and any collaborators granted Full Control permission level. So a regular collaborator trying to enter the Session will get a dialog box saying that he or she does not have access to join the Session. All collaborators will also receive an email notifying them that the Session has ended. The owner and Full Control collaborators can still enter the Session and work in it. When a Session expiration date is set, automatic email reminders will go out to all Session collaborators reminding them of the Session end date and time. Reminders are sent seven days, two days, and 24 hours before expiration. Reminder frequency is currently not customizable. The expiration date can be modified by the creator and Full Control collaborators after the Session's creation.

The new Session will open in the Studio tab and present a Session Invitation dialog box (Figure 4-6).

The Session Name and ID will be shown at the top of the box. To the left of the Invitees pane, there is a green plus sign. Click this button to invite collaborators by manually typing in their email addresses. This button is necessary if the collaborators are not in the creator's company email address book. To invite collaborators from the company email address book, click the Address Book... button below the Invitees pane. Up to 500 collaborators can participate in a Session. Use the red X button to the left of the Invitees pane to remove e-mail addresses before sending the invitation.

The Groups button to the left of the Invitee pane can be used to add large groups of users at once to the invitee list. Groups must be set up ahead of time while in a Session and can be used for all future Sessions. Groups are created by each individual Studio account holder and cannot be shared with or used by other Studio users.

When the invitee list is set, use the Message (Optional) pane at the bottom of the Session Invitation dialog box to write a custom message to the invitees (Figure 4-7). This could be something about the purpose of the Session, additional details, and so on. Session invitations

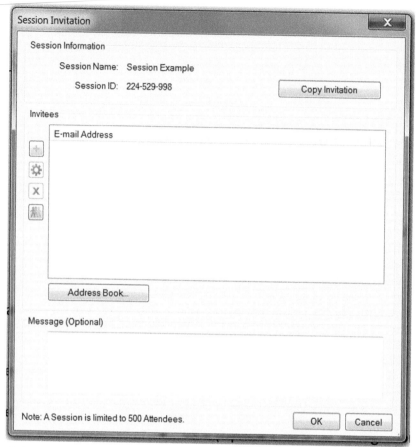

Figure 4-6: Session Invitation Window

Message (Optional)

> Hello,
> Please use this Session to review the Rev 2 P&IDs.
>
> Thanks,
> Rachel Attebery

Figure 4-7: Invitation Message

are auto-generated and sent by Bluebeam Studio, so the invitees do not have the benefit of getting the invitation from the Session creator, which can be confusing at first. It may be helpful for the creator to sign his or her own name in the Message section so invitees trust the invitation.

Invitees will receive an email from Bluebeam Studio at the address studio@bluebeamops .com. If the invitees do not seem to be receiving the invitation emails, make sure the emails are not being sent to their junk mail. If they are, have the invitees add studio@bluebeamops. com to their safe senders list. To join the Session, click on the Session ID, which is a hyperlink directly into the Session. Clicking on this link will open Bluebeam if it is not already open, log into Studio (which may require entering the user's account credentials), and open the Session.

This is a good moment to mention that every Bluebeam product—Vu, Vu iPad, Revu Standard, Revu CAD, Revu eXtreme, Revu Mac, Revu iPad—allows access to Studio. This means that users with the ability to create Sessions and Projects can invite partners, clients and vendors into Studio to collaborate without forcing them to purchase Revu.

Session Settings

Now that the Session is set up with files, default permissions, and attendees, the creator has the option to adjust more specific settings for the Session. Inside the Session, click on the Settings button at the top of the Studio tab. There are three tabs in the Sessions Settings dialog box. The first is General (Figure 4-8).

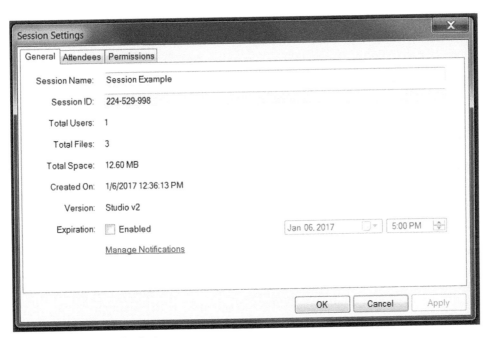

Figure 4-8: General Session Settings

The General tab shows some generic information about the Session, such as Name, ID, number of users, number of files, size of files and markups, creation date and time, Studio version, expiration date, and a link to Manage Notifications. The only pieces of editable information here are the Session Name and the Expiration date. Even if the Session creator did not activate an expiration date when creating the Session, he or she can do so here by checking the box by Enabled and choosing a date and time, and can also modify the expiration date set up during Session creation.

The Attendees tab shows which Studio accounts, by email address, have access to the Session (Figure 4-9).

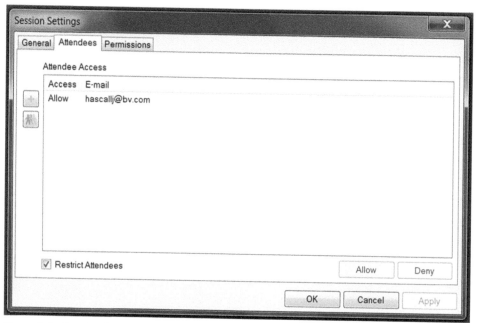

Figure 4-9: Attendees Session Settings

All the invitees invited at Session creation will be listed here. This tab allows a fast way to change the access permissions of attendees. For example, if JohnsonJ@company.com leaves the company, it is important to remove his or her access to the Session where proprietary company data is stored. To change an existing attendee's access to the Session, click on the email address in the Attendee Access pane and click Deny in the bottom right-hand corner of the dialog box. Multiple attendees can be denied at once by holding down the Shift or Ctrl key to select multiple email addresses. Attendees can be allowed to enter the Session again by clicking on their email addresses and clicking Allow in the bottom right-hand corner of the dialog box.

The Permissions tab is where the Session creator or Full Control collaborators can manage permission options for all attendees (Figure 4-10).

When first opening the Permissions tab, the creator will recognize the same permissions set during Session creation and only one group under the Users/Groups pane. These are the default permissions set for all collaborators. These permissions can be changed by clicking on the Allow/Deny dropdown next to each permission option. Many times, it is necessary to give different users different permissions based on their roles. To add a user or group with different permission levels than the default, click on the green plus sign to the left of the

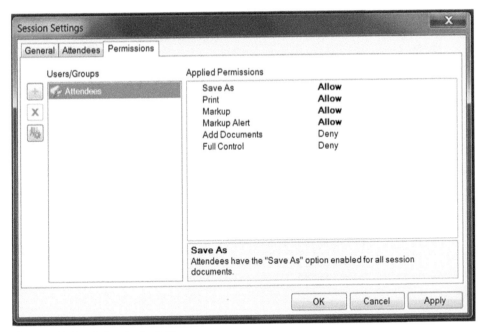

Figure 4-10: Permissions Session Settings

Users/Groups pane. The Add Users/Groups dialog box will appear, showing a list of all pre-defined groups as well as users who have joined the Session. An important note about this list is that only users who have already entered the Session at least once will appear here. A user who has been invited but has not yet entered the Session will not appear, and the creator will not be able to modify his or her permissions. This can be frustrating if the creator knows that a certain user needs different permissions than the default but doesn't want to hover over the Session waiting for him or her to join the first time. A way to get around this is to set up Groups.

To set up or modify Groups, click on the Group Management button to the left of the Users/Groups pane. This will bring up the Group Membership Management dialog box (Figure 4-11).

Only the Session creator may modify Groups; not even a Full Control collaborator will be able to modify Groups, because Groups are associated with an individual Studio account.

Unfortunately, there is not currently an option to add new members through the company email address book, or to select from the attendees already invited to the Session, and all new members must be typed in manually. Only users with existing Studio accounts can be added to Groups. Users with existing Studio accounts who have not yet been invited to the Session can still be added to Groups. The same user can be added to multiple Groups. In the case where two Groups are given different permission levels and a user is in both Groups, that user's permissions will default to the most restrictive permission level. The Session creator will be able to reuse these Groups on future Sessions.

When an Applied Permission is blank, it means it will use the same permissions as the default permissions for that function. The Full Control permission will grant a user or Group

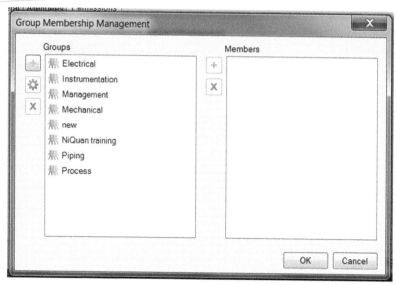

Figure 4-11: Groups Membership Management Window

the same abilities as the Session creator, except for modifying Groups. This can be helpful in many cases, such as when the Session creator is going to be out of town or just wants some help in managing the Session. It is also commonly used when the Session creator is not the Session finisher. A good example of this is when a document control or administrative professional sets up the Session with the right attendees and permissions and uploads the documents, but a lead engineer or markup consolidator has the authority to say when the review period is over and works with the documents when they are complete with redlines. Giving another collaborator Full Control allows that person to invite new users, change permissions, edit the expiration date, rename the Session, and finish the Session. When the permission levels are set up, click Apply in the bottom right-hand corner of the dialog box. Then click OK.

Working in the Session

So, now the Session creator has custom permissions and settings complete. It's time for the other collaborators to join the Session and start marking up the documents! Again, an invited collaborator can join the Session either by receiving the Session invitation email or by using the Session ID. Once inside the Session, there are a few things the collaborator should notice in the Studio tab.

The Leave button exits the current Session and returns the user to the home page where he or she can open a different Session or Project. If and only if the user has Full Control permissions, to the right of the Leave button there is the Finish button and Settings button. The Finish button closes the Session permanently, kicks out all attendees including the Session creator, and gives the option to save the marked-up files and print a report of Session activity. Finish will be covered in more detail later in this chapter.

Moving down the Studio tab, the next field is Status (Figure 4-12).

Figure 4-12: Status Bar

Status allows the user to type in any text to let others in the Session know how he or she is progressing with the review. Reviewing and Finished could be simple examples of Status. A company or group may have prescribed document review statuses, which could also be used. Once a Status is used once, it is stored in the dropdown for future use.

Under the Status field is the Attendees list. As attendees enter the Session, their names will appear in this list. If an invitee has not yet entered the Session for the first time, his or her name will not appear in this list. Next to each attendee's name, his or her Status is listed, along with the document and page number currently being viewed. If an attendee is not currently working in the Session, his or her name will appear grayed out with Offline shown next to it.

When in the Session, a user can click on the name of another in-Session user and jump to the page he or she is currently viewing. Taking this function a step further, a user can click on the orange footsteps to the far right of another active user's name to follow that user (Figure 4-13).

Figure 4-13: Follow User

A blue person image will appear behind the user being followed, letting them know that someone else is looking at what he or she is doing (Figure 4-14).

Figure 4-14: User Being Followed

The follower is now essentially viewing the screen of the followed. An orange crosshairs will appear around the mouse pointer of the followed, showing where his or her mouse is moving. As the followed makes marks or jumps to different pages or zooms in and out, the follower's screen will do the same. This can be useful during a collaborative review meeting where one user is the leader and other users need to keep up with what he or she is talking about. To stop following, the follower can click anywhere on his or her own screen.

Under the Attendees list is the Documents list. This list shows all documents uploaded to the Session. To open a document, click on its name from the Documents list. All documents can be open at the same time. When a document is opened in the Session, the Studio icon will appear in the document tab, indicating that any marks done to this document will occur in the Session and be visible to everyone else in the Session.

Under the Documents list, there are three tabs shown in the lowest Studio tab panel. The first is Record (Figure 4-15).

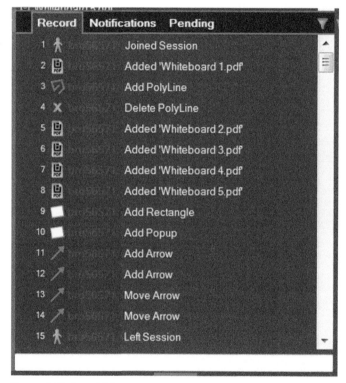

Figure 4-15: Record Tab

The Record tab shows a running trail of every action taken in the Session, in the categories of Chat, Markup, Document, Attendee, and Alert. Turn on and off which types of record are shown by clicking on the Filter button to the right of the Record tab header. By hovering the mouse pointer over any record, the user can see the name, email address, date and time, and document and page number associated with the record. Clicking on a Markup or Alert record will jump the user to the associated document and markup. While a deleted markup will no longer appear in the document's Markup Log, hiding every trace it was ever there, the markup's addition and deletion will be recorded in the Record. The Record tab can be used to chat with other Session attendees. The Record tab can be exported at any time during the Session to give a play-by-play of all Session activity. The Session finisher will also have the option to export the Record when the Session is being closed. To export the Record tab, click on the Report button to the right of the Record tab header. The Session Report dialog box will appear (Figure 4-16).

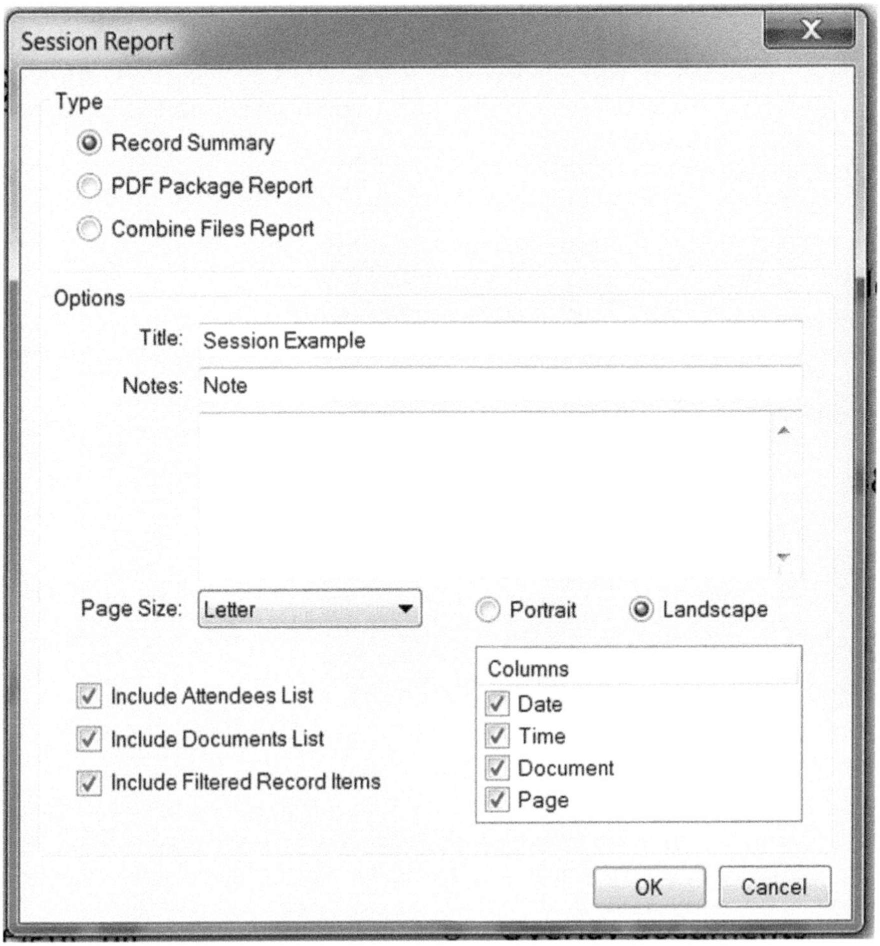

Figure 4-16: Session Report Window

Choose the Type of report. Record Summary prints a PDF list of all Record entries. PDF Package Report prints a PDF list of all Record entries and packages it with a copy of all the documents in the Session. For more information on PDF packages, see the Bluebeam Tutorial at http://bluebeam.com/us/bluebeam-university/pdf-tutorials/revu-10/pdf-packages.pdf. Combine

Files Report prints a PDF list of all Record entries and combines it into one PDF with a copy of all the documents in the Session. After choosing the Report Type, modify the Options as desired. Title is the name the report file will be saved as. Notes allows the report generator to add anything worth knowing along with the report. Choose the Page Size and orientation. Choose what information to include in the report. Include Filtered Record Items will include all Record entries, even the ones filtered out as shown in the Record tab. To include only certain entries as shown in the Record tab, uncheck this box. Choose which pieces of information to show about each Record entry: Date, Time, Document, and Page. Reports are helpful for sharing information with project team members that may not be working regularly in the Session, like managers. They can also be used to capture snapshots of the documents and their markups throughout time, or isolate specific information like attendee entrance and exit. Some projects like to use this last example to prove if all required reviewers participated in the document review.

The next tab over from Record is Notifications (Figure 4-17).

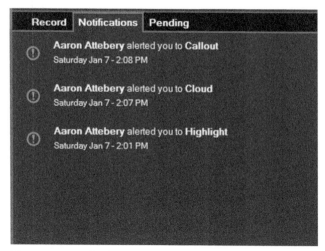

Figure 4-17: Notifications Tab

The Notifications tab keeps a list of all Alerts sent to the user by other collaborators. When an attendee is alerted of a markup, three things happen. One, an email is sent to the alerted user with information and a hyperlinked snapshot of the markup. Two, a red number icon appears next to the Session Name on the Studio home page, indicating how many new alerts are waiting for the user in that Session. Three, if the user is currently working in the Session, the Notifications tab flashes, and the alert is stored there showing the alerter, the markup type, and the date and time the alert happened. When the user clicks on the Notifications tab, new alerts will appear in orange text. Click on an alert to mark it as read. Read alerts will appear in gray text. Clicking on an alert will jump the user to the associated markup. To alert another user of a markup, the permission Markup Alert must be allowed. Once this permission is allowed, right-click the markup and click Alert Attendee, then click Choose. The Alert Attendee or Group dialog box will appear. Click on a Group or individual user to select them; select multiple Groups and users by holding down the Shift or Ctrl key. Studio summarizes all alerts from the past 10 minutes before sending an email. That way, if a user receives many alerts in a short time period, only one email will be sent to summarize them all.

The third and final tab in the lower half of the Studio tab is Pending (Figure 4-18). The Pending tab is used when a user loses internet connectivity while working in a Session.

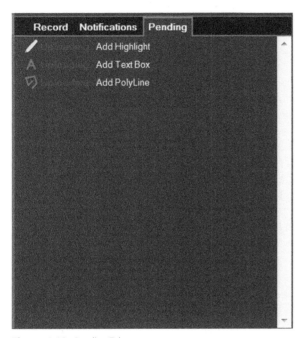

Figure 4-18: Pending Tab

All marks made while the user is connected to the internet will be automatically uploaded to the Session in real time. If the user is forced to work offline, he or she can still add markups to the documents, and they will be automatically uploaded when internet connectivity resumes. While offline, any marks made will appear in the Pending tab to let the user know that these markups have not yet been uploaded to the Session. When internet connectivity resumes and the markups are uploaded, they will automatically remove themselves one at a time from the Pending tab. If a user in the Session claims they cannot see the most recent marks made by another user, have that second user check his or her Pending tab to make sure all marks have been successfully uploaded to the Session. That concludes the explanation of Session features in the Studio tab. For more information on Studio Sessions, visit http://bluebeam.com/us/bluebeam-university/pdf-tutorials/revu-12/studio-sessions.pdf.

There are a few important restrictions to point out in Studio Sessions. First, it is quickly discovered that a user cannot edit or delete any other user's markups while the document is in the Session. Only the author of a markup can edit it. Notice the difference between a mark made by another user and a mark made by oneself (Figures 4-19 and 4-20).

Figure 4-19: Markup by Other User   **Figure 4-20:** Markup by Oneself

The yellow handles indicate the mark can be edited; the gray handles indicate it cannot be edited. This can be a perceived problem since a senior engineer will want to quickly correct errors made by junior engineers and will not be able to until the document is removed from the Session. The point of this "feature" is that with up to 500 collaborators marking up the same document in real time, it becomes very difficult to guarantee markup integrity and proper record of change, if each collaborator can edit the others' markups. In a perfect world, the senior engineer sees the incorrect mark, shakes his or her head, gets up and walks over to the junior engineer, and has a productive mentoring moment explaining why the mark is incorrect and what the correct action should be, then the junior engineer smiles and nods and changes the markup on his/her own. But alas, deadlines and time zones and geographies put teams in the less desirable position of having to fix markups without having a real conversation about them. Some options to deal with the inability to edit others' marks inside a Session are:

- Restrict markup permissions for a time to only certain collaborators. Open up the markup permissions again after the important deadline has passed.
- Use Alert Attendee to quickly notify the author of incorrect markups and request deletion or modification.
- Host a Session "cleanup" where all collaborators join the Session and a conference call simultaneously and fix markups as incorrect ones are pointed out.
- Use the statuses in the Markup Log to note which markups are incorrect. Then after the Session is finished, the consolidator can quickly filter by incorrect markups and fix them.

Custom Statuses

On a recent project, the team needed a way for lead engineers to hide unwanted markups on drawings in Bluebeam Studio. It is not possible to delete others' marks within a Studio Session, and it was undesirable to download a copy of the drawings, review, and upload again. Therefore the following solution was proposed. A custom Profile was created for the project with a custom status labeled Rejected. When a markup was statused Rejected, the mark turned white, effectively hiding it on the drawing. In order to use this status, the custom Profile had to be "embedded" in each PDF that it was used with before uploading the drawing to the Session. A Profile is embedded in a PDF when any type of change is made to the PDF while that Profile is active and then the PDF is saved. Because it is undesirable to open and modify and then save and close every PDF drawing, a batch script was written by Bluebeam Support to embed the Profile in all drawings at once. This rule applies to custom columns as well as custom statuses. Again, this must be done before the PDFs are uploaded to the Session. The following instructions outline how to create a custom Profile and status and how to embed the Profile in PDFs in a batch fashion.

Create a Custom Profile

Under the View tab, click the Profiles button.

Click Add.

Type a name for the Profile (e.g., "Jurassic"). Click OK.

Select the new Profile from the Active dropdown list. Click OK.

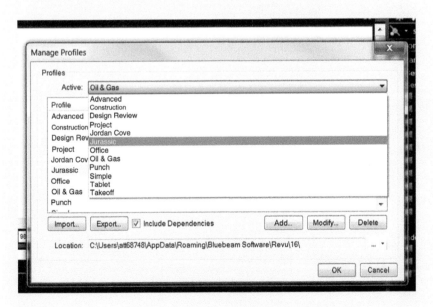

Create a Custom Status

Open up a new blank PDF by clicking the New button.

Doesn't matter what size, just use the default setting from the New dialog window. Click OK from the New dialog window to create the new PDF.

Open up the bottom flyout panel. Click the blue cogwheel Manage Status button.

In the Manage Status dialog window, on the left-hand side under Models, click Add...

Type a name for the new custom status model (e.g., "Review"). Click OK.

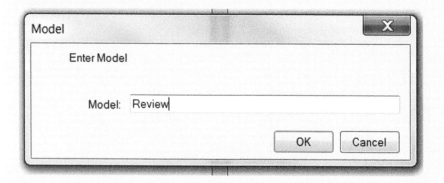

On the right-hand side under States, click Add...

In the State Properties dialog window, type the name of the custom status (e.g., "Rejected"). Check the box next to Color. Use the color picker to select the color white. Click OK.

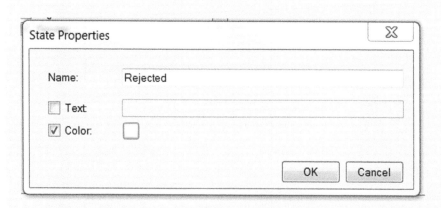

Click Save to Profile. Click OK.

Test the success of the new status by making any mark on the page. Select that mark, then click the traffic light button ("Status") to status the mark. Click the Rejected option.

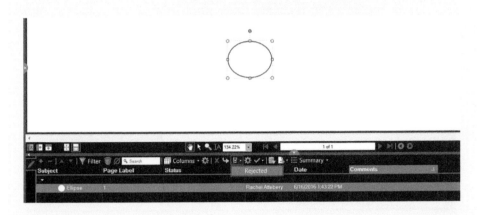

If the status has been successfully created, the markup will turn white and you will see the record of status appear in the Status column of the markup log. (If the Status column is not turned on, click the blue cogwheel Manage Columns button next to the Columns button and turn on the Status column.)

Now go to the View tab, Profiles dropdown arrow, and click Save Profile.

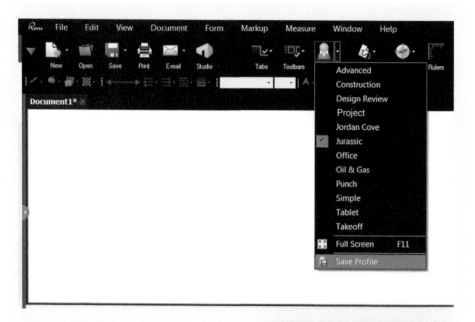

Save this PDF to anywhere on your computer (just be able to find it again!) with any name. Now you have successfully created a custom Profile with a custom status and also prepared to batch embed the Profile in the drawings.

Batch Embed Profile in Drawings

First, you will need to create the batch script. This is a one-time activity. Under the File tab, click the Batch button.

Click Script.

From the Batch: Script Manager dialog window, click Add…

Type a name for this script (e.g., "Embed Profile"). Select an icon for this script (not a critical choice).

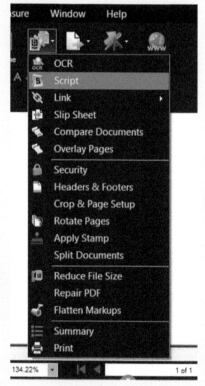

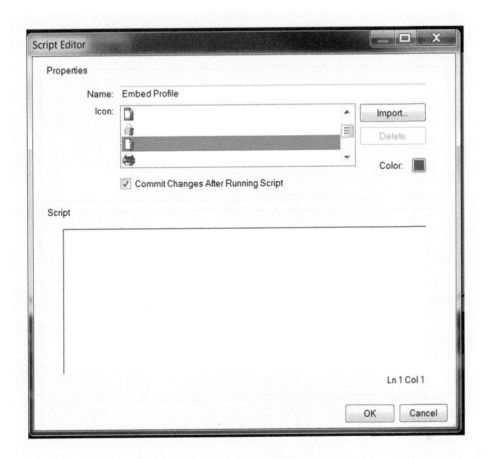

In the Script window, copy/paste the following:

```
InsertPages(0, c:\users\jadler\desktop\status.pdf, false, false,
    true, false, false)
PageDelete(1)
```

Change the highlighted text to the path of where your Profile-embedded PDF is on your computer. For example,

```
InsertPages(0, C:\Users\att68748\Desktop\JurassicProfileEmbed.pdf,
    false, false, true, false, false)
PageDelete(1)
```

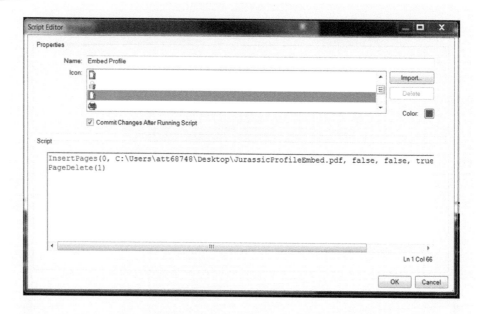

Click OK. The new batch script will appear in the Scripts list of the Batch: Script Manager dialog window.

Now you can run the script on the drawings to be uploaded to the Studio Session. With the profile embedding script selected, click Add Folder. This will let you navigate to an entire folder of drawings to run the script on all of them at once. (Alternatively, click Add Open Files to load in all PDFs you have currently open, or click Add to navigate to specific files on your computer.) Select the folder where the drawings are and click Select Folder. All the files in the selected folder will be queued for scripting. You can remove files, add files, or add another folder to the File list. Once all files are added, click Run.

This should be a very fast operation. You will not see any confirmation dialog or opening of files once the script is complete. If you want to test the success of the script before loading all the files to the project Session, open a test Session and upload one drawing. Make a mark and status the mark to ensure the Rejected status is working properly. If successful, you can now upload all the drawings to the project Session. Only the uploader of the drawings needs to run this script or even have the custom Profile with custom status on his or her instance of Bluebeam. Once the drawings with the embedded Profile are uploaded to the Session, all Session attendees will be able to use the status.

Another restriction on documents in Sessions is that they cannot be modified at the document level once uploaded to the Session. This means functions like adding or deleting pages, rotating pages, editing or deleting actual page content, redaction, digitally signing documents, and applying headers and footers cannot be performed. All these functions are available again once the document is removed from the Session. Documents of the same name cannot be uploaded to the same Session.

Finishing the Session

At the end of a review cycle, the Session can be closed and the marked up documents saved to a different location. This action is called Finish in Studio. Only a collaborator with Full Control permissions, or the Session creator, can finish the Session. For those without this permission level, the Finish button will not even appear. When a Session is finished, it is permanently closed, and no more collaborators can enter the Session to markup the documents. The Session name is removed from the Studio home page, even for the creator and Full Control collaborators, and cannot be joined again, even with the Session ID. Therefore, finishing a Session is a serious, irreversible activity and careful training should be administered to explain the difference between the Leave and Finish buttons. If a Session is accidentally finished, a new

Session will have to be created, the attendees reinvited and set up with permissions, and the documents reuploaded. This is hoping that the accidental finisher had the wherewithal to save the documents to his or her hard drive before executing the Finish function. Once a document has been removed from a Session, when it is reuploaded, any markups existing on the document at the time of upload will be locked and not able to be edited, even by the original author of the mark.

To finish a Session, click on the Finish button at the top of the Studio tab. The Finish Session dialog box will appear (Figure 4-21).

Choose which users' markups to include in the saved documents. Unchecking the box next to a user's name will permanently delete his or her markups, and they will not appear on the saved documents.

Choose how to save the documents.

- Save (Overwrite Existing) will save the documents over the top of where they came from. For example, if the PDFs were added to the Session from ProjectWise, choosing this option will save them back into ProjectWise as new versions of the files.

- Save In Folder allows the finisher to choose a folder on his or her hard drive to save the files to. This is a good choice

Figure 4-21: Finish Session Window

if the finisher needs to further edit the documents before uploading them to a repository or sending them out. Check the Session Subfolder button to save the documents to a new folder with the Session's ID.

■ Choose "Do not save files" if the documents and markups will not be needed again for anything.

Check the "Close files after finishing" box to close any currently open Session files after the Session is finished. If this box is not checked, the currently open Session files will first close then reopen from their new saved location. Under Report Options, the finisher has the choice to generate a final report from the Session's Record tab. To create a report, check the box by Generate Report, click the "…" to choose a save location for the report, and click on Settings to open up the familiar Session Report dialog box. Refer back to the section in this chapter on the Record tab for an explanation of the Session Report options. When all Finish Session options are selected, click OK at the bottom of the Finish Session dialog box. A dialog box will appear showing the health of the Finish function. If something goes wrong, the Session will not close, and the finisher must perform the necessary actions to fix the Session, or first save down the files and then deal with the Session error. Normally, the Session will finish just fine, and the dialog box will show Completed.

Click OK and watch the Studio tab jump back to the Studio home page. The Session is now completely gone and the reviewed files with their markups are saved in the designated location.

Studio Sessions for Piping and Instrumentation Diagrams

Rachel Attebery, Process Engineering, Black & Veatch

"We had a couple pockets of people in the company using Bluebeam, but my group hadn't heard of it yet. I was working on a floating liquefied natural gas project executed from multiple offices around the world, and it was a huge pain to manually maintain the paper piping and instrumentation diagrams with consistency around the world. We spent a lot of time keeping the drawings up to date. We had to scan in our markups, send them electronically to the other offices, then the other offices would print out the scans and manually transfer the markups one by one to their paper drawing set. The other offices had to do the same for us, and we had about three master sets of drawings going at the same time. Because we were using scans of the paper drawings, the hand-drawn markups didn't retain very good quality and became hard to read once printed out, and sometimes the intent of a markup was lost in translation. Not to mention, the whole process was extremely slow. We estimated about 12 engineering manhours spent each week to keep up with this paper process. The opportunity for error was higher due to the fact that three different offices were manually copying over marks from the other offices, which increased the chances of a mistake occurring or a markup getting missed. There had to be a better way.

"I was telling one of my colleagues in a different group about the horrible process one day, and he said, 'I might have a solution for you.' He showed us Bluebeam Studio, which offered a way to maintain only one master set of drawings, which would be electronic, and accessible at all times from all the offices. Markups would be made only once, and they would be instantly visible to everyone else. Making markups electronic took away the problem of unintelligible handwriting and helped improve the consistency of markups. We then created a project Tool Chest that matched the P&ID legend sheet and set up a custom Profile with specific custom columns that mimicked our normal paper management of change log for the P&IDs. We got buy-in from management by presenting on the benefits of going electronic, and didn't receive much pushback at all. The electronic process solved so many pain points that most everyone was eager to try it out. We trained all the users and created procedures for everyone to follow. Bluebeam is pretty intuitive already, so most people caught on quickly.

"One thing we did try at first and then abandon was a 55" touchscreen TV to help mimic the way people were used to putting handmarks on the drawings. We put it in a central location, just like we had done with the paper set, and expected people to walk over and mark up the drawings just like they had always done, but on an electronic surface instead of a paper surface. We thought this would help with adoption of a new process if it felt similar to the old one. Surprisingly though, the TV was almost hardly used. Once people realized how easy it was to make markups right from their desks on their own computers, no one was interested in making the trek over to the TV and dealing with the difficulty of a stylus and touch interface. The keyboard shortcuts in Bluebeam make working from a PC with the mouse much simpler than trying to get a clean markup with a touchscreen.

"We also had some trouble at first getting people to use the electronic management of change log. This wasn't incredibly different from the struggle of getting people to correctly fill out the paper management of change log. But at least in Bluebeam, any markup that was made was recorded in the markup log, regardless of whether the author completed the custom column fields or not. So we could go back and easily see who had made a markup without correctly filling out the log and have a teaching moment to correct that going forward.

"Today, the P&ID markup process is 100 percent digital. Most users really like the new system and can appreciate how much time it saves and how much more accurate the markups are. Our project was the first in our group to try this process, and since we were so successful, almost all other similar projects are using this method now. The biggest benefits I think are time saved in both dealing with all the paper and waiting to get the latest markups, better accuracy and readability of markups, more accountability since no one has an excuse not to have seen the latest drawings, and better use of resources since we're not killing half a tree anymore by printing out 255 D+ size drawings with each new version. A side benefit was that our project controls group could also go into the Session, view the design changes, and then ask us to assign a reason for the design change so they could easily track changes and how much they cost and if we needed to send a variation order notification to the client. Bluebeam made a big difference for us in all the ways that matter—time, cost, quality, and collaboration."

Figure 4-22: Start New Project

Projects

Moving on to Projects, a user can create a New Project from the Start button on the Studio home page (Figure 4-22).

The button to the right of the Leave button is Sync. Sync saves a local copy of Project files to the user's hard drive to allow viewing and editing while not connected to the internet, or "offline." To the right of the Sync button is Settings.

Project Settings

Settings in Projects looks very similar to Settings in Sessions. The General tab shows a few pieces of information about the Project. The User Access tab shows all Project attendees and if they are allowed or denied to enter the Project. The Permissions tab shows default permissions for users, named "Everyone," along with any additional users or groups with different permissions. The Applied Permissions in Projects are different than in Sessions.

- Send Invitations: Control who is allowed to invite other users to the Project.
- Manage User Access: Control who is allowed to change Allow/Deny settings in the User Access tab.
- Manage Permissions: Control who is allowed to change Applied Permissions for both the default group and for additional layers of Users/Groups permissions.
- Send PDFs to Sessions: Control who is allowed to create new Sessions from the Project and send Project PDFs to those Sessions for live review.
- Revoke Check Out: Control who is allowed to undo file check-outs done by other users.
- Share File Links: Control who is allowed to send out hyperlinks to individual Project files.
- Full Control: Control who shares equal rights with the Project creator.

The fourth tab in the Project Settings dialog box is Folder Permissions (Figure 4-23). This is an additional tab to the Settings shown in Sessions. Folder Permissions allow the Project creator to set different permission levels on the different Project folders. This is helpful since Projects can hold an unlimited number of files, and it is likely that certain collaborators should only be able to work on certain files. The five permission level options for any folder are:

- Inherit from parent: Users can do in this folder whatever is permitted for the folder one level above this folder. (The Project Root folder will not have this option, since it is the highest-level folder.)
- Hidden: Users will not see this folder listed in the Project folder structure. (The Project Root folder will not have this option.)
- Read: Users can only view files in this folder.
- Read/Write: Users can view and edit files in this folder.
- Read/Write/Delete: Users can view, edit, and completely remove files in this folder.

Click the small gray arrow next to each folder title to expand it and view folders contained within this parent folder. Every folder in the Project can be assigned a different permission

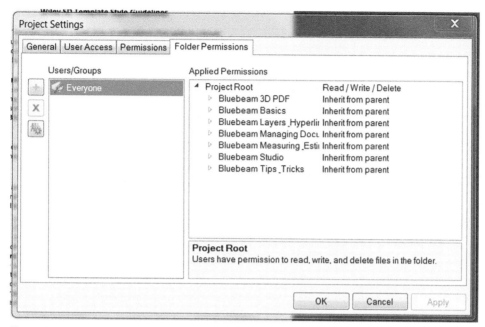

Figure 4-23: Project Folder Permissions

level. Groups set up in Projects will be available in Sessions and vice versa. Once permissions are set up or changed, click Apply then click OK.

Moving down the Studio tab, the new Project will show three options to create content (Figure 4-24).

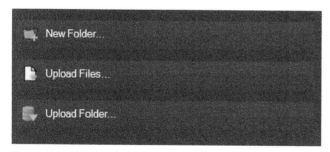

Figure 4-24: Project Add Content Options

Click New Folder to create a new folder within the Project.

Click Upload Files to choose existing files from Windows Explorer to upload to the Project. Unlike Sessions where only PDFs can be uploaded, Projects allow any file type to be uploaded. Interestingly, the uploader's computer Windows Explorer is the only option from which to select files; integrated document management systems such as SharePoint and ProjectWise are not accessible from the Upload Files dialog box. Click Upload Folder . . . to choose an entire folder from Windows Explorer to upload to the Project, same as with Upload Files, only the uploader's Windows Explorer is an option from which to select folders.

Uploading Files/Folders from Sources besides the Hard Drive

Even though the Upload Files and Upload Folder buttons in a Studio Project only bring up the uploader's computer folder structure in Windows Explorer, you can add SharePoint pages and document libraries as Favorite folders inside the Windows Explorer folder structure. This allows you to access these online sites and upload files or folders from them, even though the Studio system dialog doesn't allow you the option of choosing hard drive, SharePoint, or ProjectWise when trying to upload content. Adding a SharePoint file or folder to Studio Projects does not sync the two locations; uploading content to Projects from SharePoint creates a new copy of those files, and editing the files in Projects will not automatically save back to SharePoint. For help on adding a SharePoint site or folder to your PC folder structure, check out https://msonlinehelpdesk.zendesk.com/hc/en-us/articles/204785937-How-do-I-add-a-SharePoint-document-library-to-Favorites-in-my-Windows-Explorer-.

Working in the Project

Once folders and content are added to the Project and permission levels are set, collaborators can start working with the documents. To view a document, double-click on it. The document will open in the corresponding program, meaning PDFs will open in Bluebeam (if that is the default PDF viewer for the user's computer), Excel files will open in Excel, Word files will open in Word, and so on. Opening a file for viewing downloads a local copy of the file to the user's device, whether that is a PC, Mac, iPad, or tablet. Remember that in Studio Preferences the option Checkout on Open will cause all files to automatically be checked out to the opener upon opening. If this option is not checked, a user can check the file out by right-clicking on it in the Project file structure and selecting Check Out.

Other options available for each file are shown. Some of these options will not appear if the user does not have the corresponding permissions. For example, if the user has Read permissions only for this folder, the Check Out option will not be shown in the right-click menu. Once the file is checked out, a red checkmark will appear next to the file name. The file will also be shown at the bottom of the Studio tab under Pending Changes. This section shows all files that are checked out by the user. Files checked out by other users will not appear in this section. The user can also quickly see which files are checked out by himself or herself and which files are checked out by other users by the difference between checkout icons.

After a file is checked out, the user can edit it in its native program. Throughout the editing process, the user can update the copy saved in the Project without having to check the file back in. Right-click on the checked-out file in the Project file structure and click Update Server

Copy. This will update the Project file to match the hard copy undergoing editing, while still allowing the user to continue making changes. When the file is checked out, it is saved as a local copy on the user's hard drive. To check the file back in when editing is complete, first save the edited file in its native application. The changes will save in the local copy. Close the file. Right-click on the checked-out file in the Project and click Check In. The Check In dialog box will appear, allowing the user to enter a Comment about the new version of the file. This is a good way to keep track of why a document was edited or to note major changes that might not be obvious to all other viewers. After entering comments, click Check In. The red checkmark next to the file will disappear, indicating that the file is no longer checked out. To view the file's version history, right-click on it and click Revision History. The Revision History dialog box will appear, showing the revision number, date and time of action, type of action (Operation), email address of the user who performed the action, and comments associated with the action (Figure 4-25).

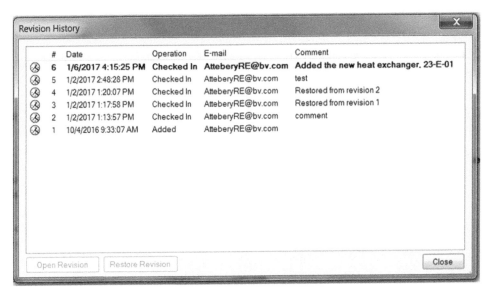

Figure 4-25: File Revision History

This dialog box can be used to view past versions of the file. To view a past version, select it from the Revision History list and click Open Revision at the bottom of the dialog box. This will download a local copy of a read-only version of the selected revision. It is also possible to restore old versions of the file, which means overwriting the current version with a past version. To do this, select the desired version from the Revision History list and click Restore Revision at the bottom of the dialog box. This will essentially check out the current version and check it back in as the restored version. The restored version will be a new revision of the file, which means the previous version is still stored in the Revision History. The new current version (the restored version) automatically gets the comment "Restored from revision #."

For PDFs checked out and edited in Bluebeam, there is a convenient icon that appears in the open file tab of the document. Opening a PDF from a Project to read-only in Bluebeam will show a small padlock symbol in the file's tab. This indicates that the file is in a read-only state and is "locked" for editing. To check out the open PDF and start editing it, click on the padlock symbol. Click Check Out. The padlock symbol will change to a red checkmark, indicating that the file is checked out and ready for editing. While the file is checked out, the following actions are available by clicking on the red checkmark:

- Check In: Check in the file as a new version with all changes to date; the red checkmark changes back to a padlock.

- Update Server Copy: Update the file version in the Project with all changes to date and keep the file checked out for further editing.

- Undo Check Out: Cancel all changes and check the file back in under the old version number; the red checkmark changes back to a padlock.

- Revert Changes: Undo all changes made since the file was checked out.

- Revision History: View the file's version history.

- Properties: View information about the file.

These actions are also available in the Project file structure by right-clicking on the file name. There are many right-click actions available in the Project file structure for each file depending on its state (checked in or out, by the user or another user, open or closed, PDF or non-PDF) and the user's permissions. See Table 4-1 for a full list of available actions. Here is a list of actions available depending on the file's state:

Always actions: Checked in or out, by user or other, Closed or Open, all file types

- Open
- Download Copy
- Share Link . . .
- Revision History
- Properties
- Copy (except Open PDF)
- Paste (except Open PDF)

Checked in, Closed or Open, all file types

- Always actions
- Check Out
- Rename (except Open PDF)
- Cut (except Open PDF)
- Delete (except Open PDF)
- Add to New Session . . . (PDF only)

Checked out by user, Closed or Open, all file types

- Always actions
- Open Project Copy
- Check In

Table 4-1: File Actions in Studio Projects

Action	Icon	Description
Open	Open	Download the most recently checked-in version of the file and open it as a local copy
Check Out	Check Out	Lock the file for editing
Download Copy	Download Copy	Download the most recently checked-in version of the file
Rename	Rename... F2	Change the name of file
Cut	Cut Ctrl+X	Cut the file from its current location and copy it to the clipboard
Copy	Copy Ctrl+C	Copy the file to the clipboard
Paste	Paste Ctrl+V	Available after selecting Cut or Copy, paste the file into a new location within the Project
Delete	Delete Del	Permanently remove the file from the Project
Share Link...	Share Link...	Create a link to download the file; see more instructions below
Revision History	Revision History	View version history of the file
Properties	Properties	View general information about the file
Add to New Session...	Add to New Session...	Create a new Session and add the file to it for live review; see more instructions below
Open Project Copy	Open Project Copy	Download the most recently checked-in version of the file and open it as a local copy
Check In	Check In	Check back in a checked-out file as a new version
Update Server Copy	Update Server Copy	Save all changes to date back to the Project file as a new version but keep the file checked out
Undo Check Out	Undo Check Out	Disregard all changes made while the file was checked out and check it back in as the old version
Replace File...	Replace File...	Replace the Project file with a new file from Windows Explorer
Revert Changes	Revert Changes	Disregard all changes made while the file was checked out but leave it checked out
Revoke Check Out	Revoke Check Out	Undo another user's checkout, check the file back in with no changes as the old version

- Update Server Copy
- Undo Check Out
- Replace File...
- Revert Changes

Checked out by other, Closed or Open, all file types

- Always actions
- Revoke Check Out

More about Share Link: This is not an invitation to the Project and will only grant the recipient of the link the ability to save a local copy of the file. This is useful for one-off scenarios where the recipient doesn't need to work in the Project but could use a copy of a specific file. The link will always download the most recently checked-in version of the file. The Share File dialog box gives the option of sharing the file for the Lifetime of the Project or for just 24 Hours. It is also possible to set a password on the link so that in case the link is leaked or shared with the wrong recipients, there is still an extra layer of security to get through before the file can be downloaded. For PDFs only, there is an option to Flatten all markups on the file. This prevents markups from being edited on the downloaded copy. The Project creator and Full Control collaborator have control over the shared links. Inside the Project, click on Settings. Go to the General tab. The Manage Shared Links button opens a webpage that shows a list of all Projects the user is involved with and their associated shared links.

More about Add to New Session: This action allows PDFs in the Project to be added to a new Session for a live review. Adding PDFs to a Session will check them out in the Project; when the Session is closed, the marked-up PDFs are checked back into the Project as a new version. To send PDFs from a Project to a Session, right-click on any checked-in PDFs and click Add to New Session... If a user wants to join the Session from the Project but has not yet been invited to the Session, the user can right-click on the file name and click Join "Project Name"—Project ID. The creator of the Session will be sent a notification letting him or her know that someone has requested to join the Session, and the creator can then Allow or Deny the requestor access to the Session. The PDFs sent to a Session are checked out and cannot be checked out by another user while they are in the Session. Users with the appropriate permissions can Revoke Check Out by right-clicking on the file name and selecting Revoke Check Out. This will leave the PDFs in the Session, but also cancel the Project checkout and disregard any markups made in the Session. When the live review is finished, the Session creator or Full Control collaborators can click Finish to close out the Session. The familiar Finish Session dialog box appears. The Session finisher still has the option to save the PDFs to a local folder or not to save the files, but to save the files back to the originating Project, choose the Save (Overwrite Existing) option. Continue editing the PDF as necessary, then click the red checkmark and select Check In to check the marked up PDF back into the Project. If a Session has already been launched from the Project and is still open, Project PDFs can be sent to either a new Session or the existing Session.

Another action available, depending on the user's permissions, by right-clicking in the blank Project space is Show Checkout Review (Figure 4-26).

This will open the Checkout Review dialog box, which shows all currently checked-out files in the Project and who has checked them out. If a PDF file has been added to a Session

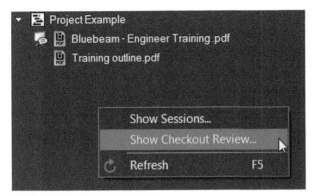

Figure 4-26: Show Checkout Review

from the Project, the Session Name and ID will appear in this window as well. Users with Revoke Check Out permissions can undo checkouts from this window by selecting the files to be checked back in and then clicking Undo Check Out. Select multiple files at once by holding down Shift or Ctrl before clicking Undo Check Out. Click Refresh to get the latest checkout information.

Besides actions available for individual files, there are folder actions available by right-clicking on Project folders. The actions available are dependent on the user's permissions.

- New Folder: Add a new folder underneath the selected folder.
- Upload Files: Add files to the selected folder from Windows Explorer.
- Upload Folder: Add an entire folder underneath the selected folder from Windows Explorer.
- Download Copy: Download a local copy of the entire selected folder.
- Rename: Rename the folder.
- Cut: Cut the folder and copy it to the clipboard.
- Copy: Copy the folder to the clipboard.
- Paste: Paste the folder from the clipboard to a new location in the Project (available after selecting Cut or Copy).
- Delete: Delete the entire folder and all its contents.
- Show Checkout Review: View all checked-out files contained within the selected folder
- Properties: View general information about the folder.

EXPERT TIP

Using Project Path for Hyperlinks

It's sometimes helpful to provide a hyperlink to a Studio Project file. For example, an engineer working on a project where a dashboard was set up using SharePoint as the main home page and a Studio Project with hyperlinked PDFs as a subportion of the main dashboard. She tried just inviting people into the Project, but they got confused about what page to start on and forgot about this part of the dashboard since it wasn't linked into the main SharePoint page. To fix this, she created a link from the SharePoint page directly to the main page of the Project dashboard. Inside the Properties of a Project file, the Project Path is also a hyperlink to that file.

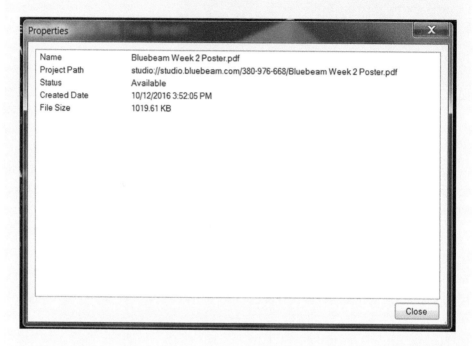

Right-click, Copy to paste the link to the clipboard, then paste it somewhere to help someone get into the right starting point of the Project. Make sure each of the users are added to the Project under User Access in Settings. They don't even have to know they have been added. All they need to know is what to click on from the SharePoint page, and that link will open Bluebeam, open the Project, and open the PDF file in Bluebeam to get them started on the right page.

By right-clicking on the project root itself, more Project actions are available, depending on the user's permissions:

- New Folder: Add a new folder to the Project.
- Upload Files: Upload files to the Project from Windows Explorer.
- Upload Folder: Upload an entire folder to the Project from Windows Explorer.
- Download Copy: Download a local copy of the entire Project.
- Show Sessions: View all open Sessions launched from the Project.

- Show Checkout Review: View all checked-out files in the Project.
- Refresh: Update the Project to get the latest information.

Working Offline in Projects

A huge benefit of Studio Projects is the ability to work offline. This feature is used often by field personnel who have poor connectivity outside the job trailer. Working offline can occur after the user syncs a copy of the Project files to his or her device to view or edit them even without internet connectivity. After regaining connectivity, the user can check in the edited documents and resync to get the latest version of all files. To work offline, first choose which files and folders should be synced to the local device by selecting the desired file/folder and clicking the green Sync Include icon to the right of the file/folder name. This marks that file/folder for download upon clicking Sync at the top of the Project. Having the option to choose what content to sync decreases download time and saves storage space on the user's device. The user can turn on/off the files/folders for sync at any time. For example, he or she might be working on one set of files today and another set of files tomorrow. After working today, he or she can turn off sync on the completed files and turn it on for the files to be used tomorrow. The user can tell which files/folders have Sync Include turned on by looking for the green icon to the right of each file/folder. After selecting the files/folders to be synced, click the Sync button at the top of the Project.

Now click the dropdown arrow next to the plug icon and select Work Offline. Any files that are still shown clickable have been synced and can be viewed or edited while working offline. If a user knows he or she will be editing certain files during the day while offline, it's a good idea to check out these files before going offline. That way, the user locks these files for editing, so others cannot edit the files, and the user can make all necessary changes while offline. When connectivity is resumed, the user can check these files back into the Project with all new changes and be assured that there are no conflicts between his or her new version and someone else's who was also editing the checked-in file during the day. If the files are not checked out before going offline, the user can still edit them offline by using the Edit Offline function. Check the file back in to add the offline changes into the Project as a new version. If another user has checked out or edited the file, a yellow Conflict icon will appear next to the file. If this happens, contact the Project creator or appropriate person to resolve any conflicting changes. It is typically best practice to avoid editing offline if at all possible, since it compromises transparency in the document record and creates multiple sources of truth.

Closing the Project

Projects will remain open indefinitely and do not have to be finished like Sessions. To close a Project, find the Project in the Projects list on the Studio home page. Right-click the Project name and select Delete. This will delete the Project and all files and folders and will not save a local copy. To save a Project before deleting it, download a copy of the entire Project first, then go to the home page and delete it. If a user is done working in a Project but doesn't want to delete it, or doesn't have the permissions to delete it, right-click on the Project name in the Studio home page and click Remove from List. The Project will no longer appear on the user's Studio home page.

Managing Notifications

Because a user may have many active Studio Sessions and Projects, it is helpful to know how to customize Studio notifications, to limit emails only to the desired pieces of information. To do this, go to the Studio section of Preferences by clicking Settings on the Studio home page and click Notification Preferences (Figure 4-27).

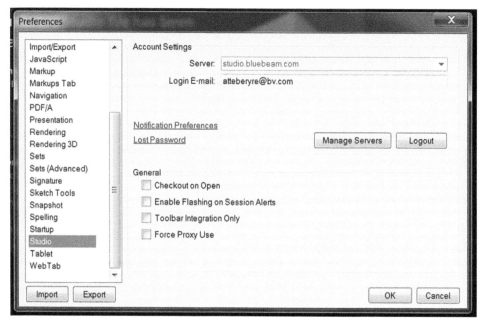

Figure 4-27: Studio Notification Preferences

Conclusion

In conclusion, Studio is a powerful feature that can be used by anyone with an internet connection and some form of Bluebeam installed. Studio Projects is the one portion of Bluebeam that allows users to interact with non-PDF files. Studio Sessions is quite possibly the best PDF live collaboration tool available today. By combining Projects and Sessions, users can create a common data environment for many collaborators with a single source of truth, live markups, change management, and the ability to work either online or offline, in the office or in the field.

Chapter 5
Management of Change

Over the last 5 years, perhaps the most exciting changes to Bluebeam Revu have been the added features and changes made in the way of document management. Instead of being solely a markup tool, Revu is growing into a powerful construction document management tool.

Three of the biggest additions are Digital Slip Sheeting, Bluebeam Sets, and Tags, all of which will be covered in this chapter. Specific topics include:

- Digital Slip Sheeting
  - Overview
  - Matching documents
  - Transferring markups
- Bluebeam Sets
  - Overview
  - Creating a set
  - Updating, modifying, and maintaining a set
- Tags
  - Overview
  - Adding and customizing tags
  - Utilizing tags

Digital Slip Sheeting

It's often said that change is the only constant in design, and as design engineers, the authors cannot agree more. Design projects are always in a continuous state of flux. Models, ideas, and drawings get updated on a weekly, daily, or even hourly basis, making it difficult for human professionals to stay abreast of everything that's going on. Luckily, computers are very good at organizing, tracking, and evaluating differences in documents and can help maintain order from the chaos.

For years, professionals have utilized Track Changes in Microsoft Word to collaborate on documents, monitoring and understanding changes others have made without memorizing the entire document or reviewing it in full each time it's opened. Unfortunately, Autodesk AutoCAD, Autodesk Revit, and other design tools don't explicitly have a feature equivalent to that. They also don't have the high-powered markup capabilities discussed in chapter 3, making the design software difficult to use as a revision or markup tracking tool.

Bluebeam recognized that gap and made strides to fill it with the introduction of the Digital Slip Sheet process, a process by which markups from a previous version of a document are copied to the current version of a document.

Overview

Throughout their lifetime, all construction drawings and documents go through a number of revisions. Prior to being issued for construction, those revisions may or may not be managed or tracked by the designers, and after being issued for construction they are almost always managed or tracked vigilantly.

As an example, consider a generic project. A typical workflow might include an engineer making markups or redlines on an original PDF drawing (Revision A). Those redlines might be the result of that engineer's calculations, a change dictated by the client, or a change driven by another engineering discipline. In any case, the redlines relay the changes to the designer, who then updates the model and re-creates the associated, updated PDF drawings (Revision B).

Unfortunately for the designer, the redlines don't stop flowing while he or she updates the drawings. Instead, they continue to be added in parallel with the revisions. As such, it is nearly impossible to finish all the updates at once.

From the design engineer's perspective, there is a critical point in time where it becomes difficult to continue to leave markups because the Revision A drawings do not yet reflect changes he or she knows exist.

The designer wants to incorporate all the markups before providing an update, but the engineer wants the update as soon as possible. Their interests are conflicting and the result is an exhausting effort of tracking both revisions and markups to make sure that everything gets incorporated correctly.

Any professional who has been responsible for managing revisions can attest to the fact that manually documenting revisions with a spreadsheet or other method is labor intensive and prone to human error. Further, transferring and tracking markups is repetitive and mundane. With the addition of Digital Slip Sheeting, Bluebeam simplified the revision process, freeing up the workforce by enabling the computer to manage mundane tasks, and create a quick and easy way to differentiate between new drawings and old drawings.

The Digital Slip Sheeting process within Revu automatically handles a few of these tasks for the designer and the engineer, simplifying the workflow and reducing the human effort significantly. The main features and capabilities of Digital Slip Sheeting are as follows:

- Identifies and matches pairs of drawing sheets (Revision A matched with Revision B).

- Applies a PDF "SUPERSEDED" stamp on the original (Revision A) drawing sheet, identifying it as being replaced.

- Collates the combined (new and original) drawing sheets in the appropriate order, maintaining the original (Revision A) drawing and markups for reference.

- Transfers markups from the original drawing set (Revision A) to the new drawing set (Revision B), placing them exactly in the same location on exactly the same sheet.

With Digital Slip Sheeting, Bluebeam saves time, saves frustration, and generally simplifies the job for the project team, allowing more time to focus on the important aspects of design and generally do more with less.

Matching Documents

Digital Slip Sheeting begins with identifying the Current Pages (Revision A) and the Revised Pages (Revision B) so Revu can identify the matches. To begin, the user should click File > Batch > Slip Sheet. The window shown in Figure 5-1 should appear where the user can select the files of interest.

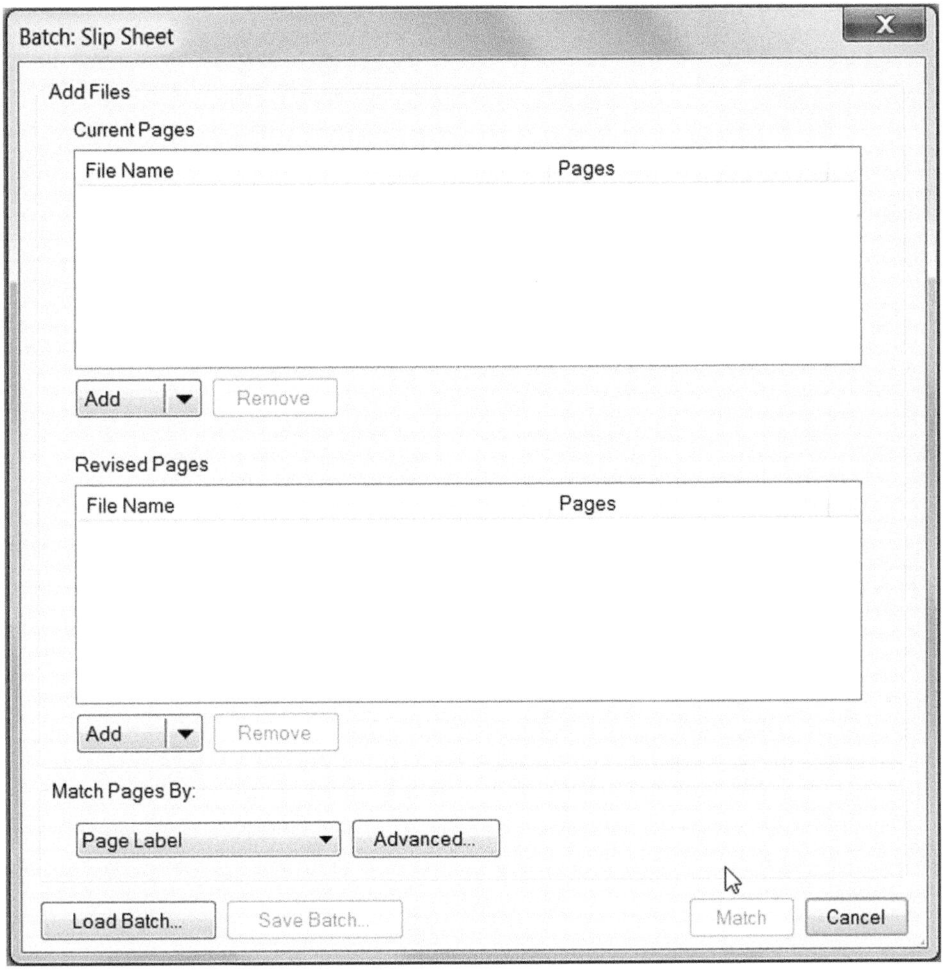

Figure 5-1: Batch: Slip Sheet Window

Just below the Current Pages section of the window, click "Add" or select the down arrow on the right side of the button to display additional options. After choosing the most appropriate option, the user should navigate to the location of the original (Revision A) files or folder and select them. The files should then be listed in the Current Pages window, as shown in Figure 5-2.

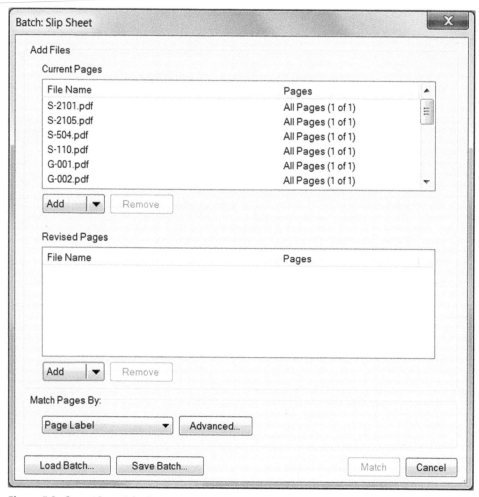

Figure 5-2: Current Pages Selection

The user may then repeat the process for the Revised Pages window, identifying the new (Revision B) drawings, which will again populate a list of files within the pane, as shown in Figure 5-3.

The final step in setting up the match is perhaps the most important. In order to identify matching pairs, Revu needs to know what piece of data should be used to attempt the match. Various options exist, including File Name, Page Label, and Page Region. A manual option also exists, but matching manually negates the benefits of the tools and should only be used in cases where a more automated match is unsuccessful.

- File Name: Revu looks for matching file names. The user should use this option when consistent file naming is used between new and original files.

- Page Label: Revu ignores the file names and looks for matching page labels. When relevant labels exist, for each sheet, Page Label is an excellent method because the order

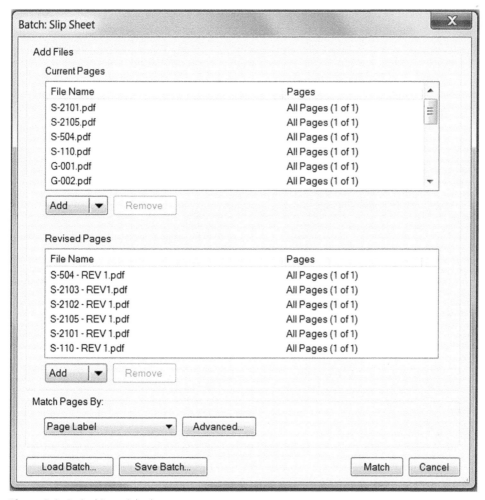

Figure 5-3: Revised Pages Selection

of the sheets in a multipage file does not influence the matching process. It does in the File Name option.

- Page Region: Revu allows the user to indicate a boundary in which to look for content on each sheet. The software then extracts the content of that boundary, converts it to string data, and uses it to compare to other sheets. This is an exceptionally handy alternative when relevant page labels do not exist, instead being labeled 1, 2, 3, and so on. An example of this would be the placement of the boundary around the drawing sheet number (G-100). Bluebeam would then extract G-100 and find its match.

Once the files have been selected and appear in the list, the user may click Match in the bottom right. A new window, similar to that shown in Figure 5-4, will appear with two tabs at the top. On the left tab, Matched Pages, Revu indicates the pages it believes to be a match. The Current Pages (Revision A) show up on the left and the Revised Pages (Revision B) show up on the right. Each row represents one match and the user may adjust any of the matches as necessary using the up/down arrows and the remove (x) button. The second tab indicates a list of sheets that were not matched. In some cases there may not be a matching revision for an existing current page. In other cases, the matching data, page label/file name/and so on didn't match. NOTE: The syntax of the matching variables is highly important. If one variable utilizes spaces before and after a hyphen and the other does not, Revu will not identify the match.

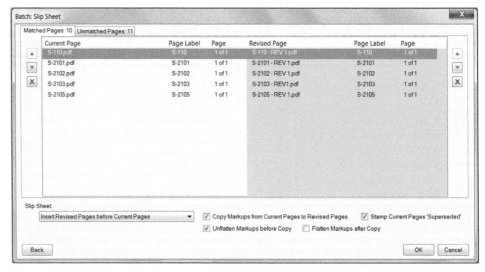

Figure 5-4: Matched Pages Window

After adjusting the matches as needed, the user can choose a number of options for the output, such as choosing to insert the revised pages (Revision B) into the original (Revision A) file(s) with each revised sheet appearing before the matching original sheet, or choosing to fully replace the original (Revision A) sheets with the revised (Revision B) sheets, eliminating the corresponding Revision A sheets completely.

EXPERT TIP

Advanced Features

Recognizing the enormous variety of workflows, Revu developers kindly built in several conveniences to aid in successfully matching drawing sheets.

The first is the simple idea that it's possible to match sheets that are part of a multi-page file with sheets that are stand-alone files. Using the page label matching technique, Revu will ignore the file structure completely. What this means is that the user doesn't have to reproduce the whole set of drawings in order to update the handful with revisions. As one, two, or five sheets are revised, they can be run through the Digital Slip Sheet process and incorporated into a larger set of sheets without hassle.

The second convenience is the ability to filter the match terms. Although a human could quickly realize that sheet A-500 (Revision A) and sheet A-500-Rev B (Revision B) are the same, the computer recognizes these file names as being different due to the "-Rev B" addition on the second file. To aid in the matching, Bluebeam is capable of applying filters to the matching terms. From the Digital Slip Sheet window shown above in Figure 5-1, the user may choose Advanced . . . and be presented with the supplemental window shown in Figure 5-5. The window contains an entry box where the user may identify the desired filter and a list of wildcard characters representing various patterns of characters and numbers. In the example above, a filter of A-# could be used to tell Revu to look for the matching 500, and ignore the rest. The feature may seem insignificant, but it enables a wide variety of naming convention and, in the case of matching file names, allows multiple versions of a file to be given a unique file name and be placed in the same folder.

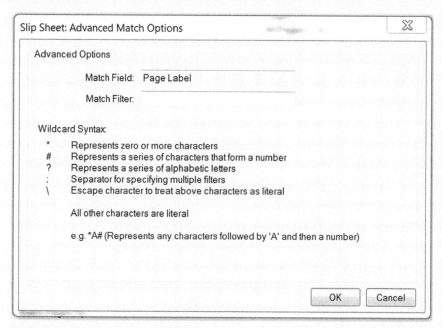

Figure 5-5: Advanced Match Options Window

The third convenience is the ability to save and reload a batch. Often in the life of a project, the files being revised are the same, just new revisions (Revision C). Instead of having to set up the match each time, the user can Save Batch . . . and Bluebeam will store the matching setup on the local computer, allowing it to be recalled the next time revisions are updated. The batch files are standalone files and may be stored anywhere at the user's preference. To recall a batch, simply click Load Batch . . . and navigate to the saved batch file.

Depending on the option chosen, the user has supplemental checkbox options, which should be explored independently. Each checkbox represents an automated process that has the ability to reduce the effort required to manage drawings.

- The Copy Markups option transfers markups or redlines from the original to the revised in the exact location those markups originally appeared.

- The Stamp option provides a means of clearly indicating which drawing is old and which is new to reduce confusion within the design and construction team.

- The two flattening options allow the extraction and placement of markups from the PDF content layer.

EXPERT TIP

The Markup Layer and the Content Layer

In the realm of PDF files, there are two main layers, the markup layer and the content layer. An easy way to think of these is to consider that the content layer is where everything from a CAD or word processing software is placed. When one creates a PDF from Word, the text is on the content layer. The same is true with AutoCAD, Revit, and others. The line work, details, sections, and so on are placed on the content layer. Markups, on the contrary, are placed on the markup layer, almost floating above the content layer.

Revu enables the user to move markups from the markup layer to the content layer through a process called "Flattening," identified by the yellow steamroller icon, 🛞. Flattening can be completed on a single markup, a selection of markups, or all markups through various options and methods. Unflattening, by contrast, is of course the moving of the markups from the content layer back to the markup layer, the reverse of flattening.

The flattening and unflattening options mentioned above relate to the transfer of flattened markups from the original to the revised. Markups that have been moved to the content layer by flattening will not be transferred unless they are first unflattened.

The selection choices on those two boxes will be important to the user and will depend on the specific scenario desired.

Clicking OK will bring the user to the Unmatched Pages window where he or she will be able to decide what should be done with the pages that were not explicitly mapped. Revu conveniently allows the user to extract the unmatched pages individually and open them in order to then decide what to do with them. If the unmatched pages were expected, selecting Skip will bring the user to the final menu, which summarizes the process and enables the user to create a summary report. Upon clicking OK, the program will either create the

Slip Sheet Summary and open that summary file or return the user to the main Bluebeam window. Upon opening the newly saved matched file, the user will see the changes he or she selected. A sample is provided in Figure 5-6.

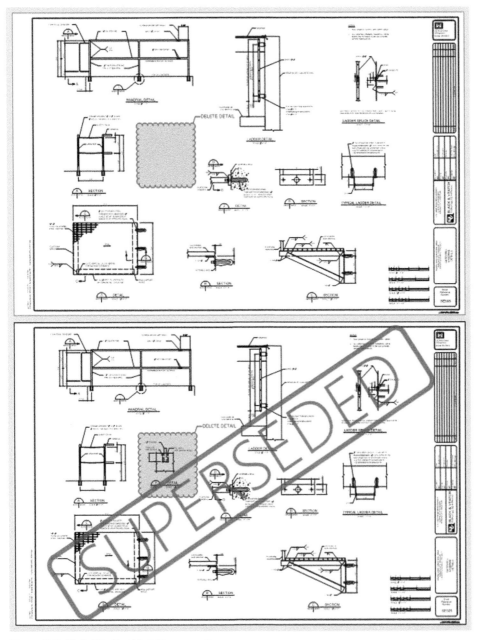

Figure 5-6: Digital Slip Sheet Results

Digital Slip Sheeting Is Difficult to Undo

Computers have brought enormous simplifications for human users. Unlike with a type-writer, a misspelled word can simply be deleted and retyped rather than covered with white-out and retyped, leaving a permanent mark on the document. "Undo" capabilities built into nearly all the software packages of today have changed our workflows and offered incredible time saving.

The Digital Slip Sheet process, however, is capable of making saved changes to hundreds of drawings at once. As soon as those changes are saved and committed to the files, the Undo feature is no longer relevant and reversing changes must be done manually. The truth is, Digital Slip Sheeting is very difficult to undo.

A reader or user wanting to set up a set of demonstration files to use for multiple demon-strations should keep this in mind and reserve a master copy that remains untouched by the Slip Sheet process. Additionally, it is extremely important to verify the matched files before starting the process, to avoid a difficult situation.

For more on the Digital Slip Sheet process, users may find a descriptive video on Bluebeam's website at www.bluebeam.com/us/products/revu/batch-slip-sheet.asp.

Bluebeam Sets

As daily users of Bluebeam, the addition of the sets functionality is one of the authors' favor-ite upgrades. It takes into account typical and simple workflows from CAD software pack-ages, allows Revu to do some heavy lifting, and significantly improves the functionality of a drawing set for the user.

For decades, designers have used two competing philosophies when creating CAD draw-ings: store all the sheets in one project file, or store all the sheets in individual files. When using the first philosophy, publishing sets of drawings is fairly simple, but other problems ensue such as large file size and multi-user sharing. For both of those issues, individual files is the simpler philosophy, but publishing drawing sets requires the extra step of combining documents, which opens an entirely new error source.

Bluebeam Sets functionality offers the best of both worlds without introducing human errors.

Overview

A Bluebeam Set allows the user to scroll through and use a series of individual PDF files as if they were one single combined file. The Set or .bex file is a special type of file that does not include any PDF content, but rather maintains relationships between other PDF files. As such, it is a very small file, typically less than 100 kB.

The set functionality is wonderful because it allows designers to continue to work in individual CAD files per sheet, without burdening them with the hassle of combining the files into a multi-page document for the engineers or architects to use. Sets offer the best of both worlds. Further, managing changes and revisions in a set is easy because the Digital Slip Sheet process is built directly into the feature.

An example of a Bluebeam Set can be seen in Figure 5-7. The thumbnails of the files within a set are displayed in the Sets tab that can be located in any of the three panels on the left, right, or bottom sides of the screen. In this particular case, each of the thumbnails represents an individual PDF file containing exactly one sheet. The set file manages the order, organization, and display of these individual files so they look and scroll as if they are one multi-page file, something architects and engineers definitely prefer.

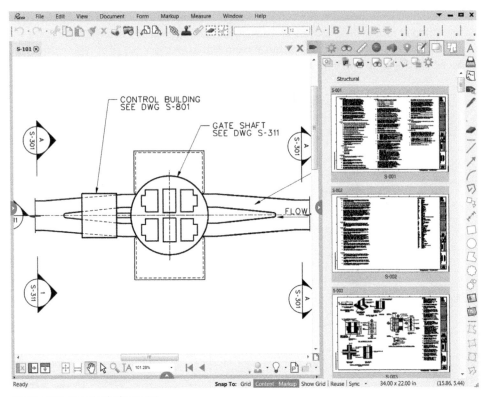

Figure 5-7: Generic Bluebeam Set

Clicking on one of the thumbnails will open that single page file in the main window for review and markup. Clicking on a second file will close the first file and open the second file in its place. A very nice feature of this workflow is that the only files a user has open are the ones appearing in the main window, leaving all the other files (pages) available for use by other members of the project team. This eliminates the read-only issue that occurs when working on large projects with large combined files.

Curious readers will be wondering what happens when an individual file from a set is opened and modified. What happens to those changes when the second file is opened before the first file is saved? The answer is nothing. Revu recognizes that the first file has unsaved changes, but doesn't want to decide for the user whether those should be saved, so it simply opens the second file in a new file tab, leaving the first file open and unsaved, available for the user to decide what to do with it. The one caveat is that the user now has two files open, preventing both from being accessed by others.

EXPERT TIP

The Open File Stockpile

As shown in Figure 5-8, working with sets is so easy that many times the user hasn't even realized that he or she has modified tens of pages without saving a single one, resulting in a string of tabs with an asterisk on each. Aside from the risk of a data loss, this isn't a significant problem for Revu or the user at hand, but it does lock out all other users from accessing the file in anything other than read-only mode.

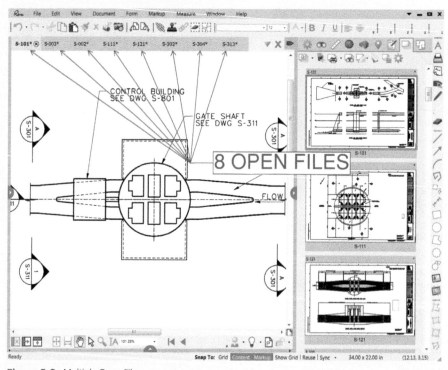

Figure 5-8: Multiple Open Files

One of the beautiful things about working with sets is that multiple professionals can be working in the files simultaneously so long as they aren't accessing the same individual page. Considering a drawing set of 100 sheets, it's highly unlikely that users will need to spend significant amounts of time on the same sheet at exactly the same time, thus nearly eliminating the read-only issue that would wreak havoc when using a multi-page file of 100 sheets. It is not the real-time collaboration functionality of Bluebeam Studio, but it is a very workable scenario.

As such, it becomes important for users to understand what they are doing, recognize the impact it has on other members of the project team, and minimize the number of unsaved files he or she has open.

Although the example above noted the case where each sheet was a single file, a beneficial feature of the Sets functionality is that Revu doesn't care whether files are single pages or multiple pages. It will manage both and any combination of the two. The one catch is that when a user opens a sheet from a multi-page file, he or she has simultaneously opened all the other sheets in that file and prevented other users from accessing them at the same time. Though the authors do recommend using single sheet files, utilizing multi-sheet files isn't a significant problem, but can sometimes be a source of frustration within a project team. Users simply need to be cognizant of what they have open and what impacts that has on others.

Highly observant readers will have noticed from Figure 5-7 that the set file also categorized the drawings by discipline, General, Structural, and so on. The categories are user-defined and based on the letter at the front of the sheet name. S-500, for example, might be identified as a Structural sheet because "S" had been identified as "Structural" in the category setup. The same could be said for sheet G-100, which might be categorized as "General." As noted, categories are completely user customizable and will be discussed later in this chapter.

The final benefit of sets that will be discussed here is the ability to "stack" revisions. Revu developers recognized that users often want to see what was and compare it to what is, so they built in the revision stack. Looking closely at an individual sheet, as shown in Figure 5-9, one can see a small blue arrow in the bottom left corner and a ghost sheet hiding to the left side of the sheet. Clicking that arrow will reveal the previous revision of the sheet hiding below, often stamped superseded and always covered by a red X, indicating the sheet is not current. Revu is capable of storing unlimited revisions of a single sheet and sets allows the user to browse through those easily, backward or forward. Should a user want to open a previous revision, he or she should simply click on the thumbnail just like the current revisions.

For more on sets, check out this great video provided by Bluebeam on their website: http://bluebeam.com/us/bluebeam-university/training-materials/sets.asp.

Creating a Set

Creating a set is simple, but before doing so the authors recommend that the reader prepare the sheets to be included as follows:

- Create each sheet to be included in the set in PDF format. This can be done using the native plot functionality of the CAD software, or through the Bluebeam Add-In Create PDF functionality.

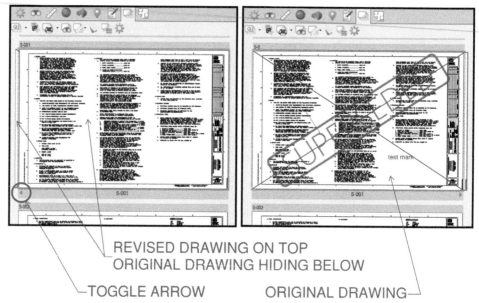

REVISED DRAWING ON TOP
ORIGINAL DRAWING HIDING BELOW

TOGGLE ARROW   ORIGINAL DRAWING

Figure 5-9: Stacked Revisions

- Plot each individual sheet at the same sheet size with the same settings. This ensures the border is identically placed on each file and will be beneficial to the process.

- Plot each individual sheet with vector content. Vector content produces higher-quality imagery with smaller file sizes. It allows the user to zoom in indefinitely without losing clarity and also allows for the creation of searchable text.

- Locate the sheet files in one central repository, all in the same folder or organized by category into multiple subfolders.

Although those preparation steps aren't necessarily required, they will help the user get the best results in the final product.

Within Revu, open the Sets tab by clicking the 🗋 icon or typing Alt+2. In the top left corner of the Sets tab, the user will see the Modify Set icon 🔯 ▾ with an adjacent down arrow. Clicking the down arrow reveals multiple choices including New Set, Open Set, and Close Set. Click the New Set option 🗋 New Set .

The window shown in Figure 5-10 should appear. Here the user has the ability to add files to a set using the Add button or the adjacent dropdown. Similar to the file match window from the Digital Slip Sheet Process, files may be added by navigating to and selecting individual files, selecting currently open files, or selecting a folder that houses all the desired files.

Once the selection is made, the user will be presented with the Categories Template window, as shown in Figure 5-11. Here the user may select from Bluebeam's default "Construction" template, existing custom templates, or create his/her own template by clicking the "Templates" button. The authors have found Bluebeam's Construction template, which utilizes AIA standard naming convention, to be very versatile and encompassing.

Figure 5-10: New Set Window

Figure 5-11: Categories Template Window

For cases where the default will not be sufficient, users can modify an existing template or build their own from scratch in the Manage Templates window, as shown in Figure 5-12. Here users can modify, add, or delete category names, assign one or more identifying filters to each category, and save the template for use on future projects under a user-defined name.

After the appropriate template has been selected, continue with file or folder selection and return to the New Set window shown in Figure 5-10 above. Before proceeding, the user has some important selections to make. First and foremost, the user needs to decide whether or not to use relative paths for the .bex Set file. With relative paths, if the Set file and individual files are moved together to a different location, the set will still function. Without relative

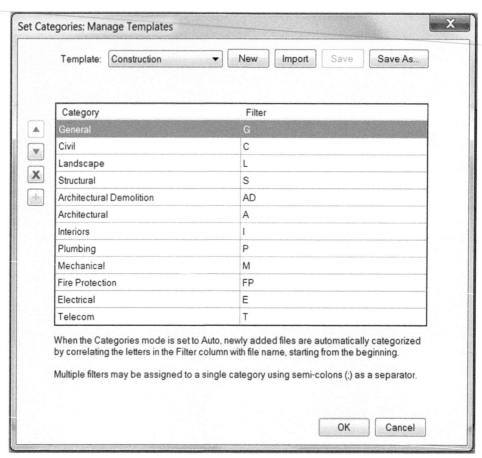

Figure 5-12: Manage Templates Window

paths, the same scenario would result in a broken Set file because the links are still pointing to the original file location, where the files no longer exist. On the contrary, if the .bex Set file were to move on its own, relative paths would cause the file to break and non-relative paths would not.

The next set of choices and settings can be found under the Options button, which launches the Set Options window, as shown in Figure 5-13. Here the user can adjust the display and sorting of the files within a set. As an aside, the sorting of the categories must be done in the template setup or the Categories tab.

Beyond the display options, the reader should recognize the remaining selections as being very similar to the matching options from the beginning of this chapter. Anticipating revision postscripts such as "- Rev B" or "-Rev 2" between subsequent file names, Revu has a built in Revision Filter that will automatically identify and process simple naming conventions like those noted. For more complex naming schemes the Wildcard Syntax can once again

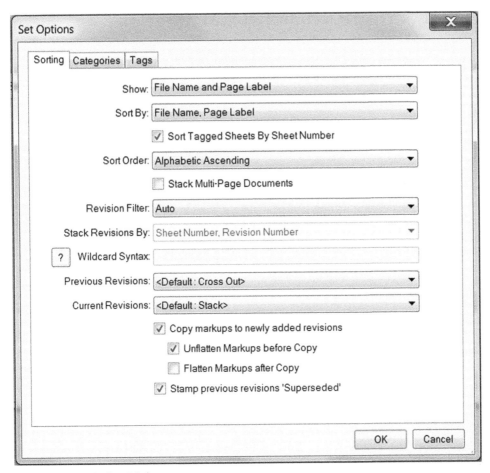

Figure 5-13: Set Options Window

be employed to correctly match drawing sheets. Finally, at the bottom of the window, the user has the options to transfer markups forward and stamp previous sheets as Superseded.

There are two additional tabs in the window, Categories and Tags. Categories have been discussed previously in this chapter, but can be created or modified in this window also. Tags are a powerful newer feature and will be discussed at the end of this chapter.

After modifying the settings as desired, click OK twice. The user will immediately be prompted by the Add Tags window. As noted, tags will be discussed in detail at the end of this chapter. The reader may want to read on to learn more about tags or skip the tags for now by selecting the dropdown on the bottom right of the window and choosing "Skip Tag" for each pop-up window. The final Tags window may be closed.

At this point, a set has been created and should be displayed in the Sets tab as thumbnails of the drawings. The .bex Set file does not save by default and therefore should now be saved using the Modify Set dropdown menu.

Updating, Modifying, and Maintaining a Set

As noted at the beginning of this chapter, change is constant. Bluebeam developers realize that and enabled the .bex Set files to be changing, working files so they could always be kept up to date with the project.

Adding Sheets

The first and most important modification to a set is adding sheets. Throughout the design and construction lifecycle, the addition of new revisions of previous versions and brand new sheets is likely inevitable, so both need to be accommodated.

Within the Sets tab, the user can click the Add Files to Set icon, 🔳 A Windows Explorer window will pop-up and the user will have the ability to navigate to the desired files, select them, and click Open. This will add files to the set, matching, stacking, and handling them as setup in the creation of the original .bex Set file. New revisions of old drawings should be stacked; brand new drawings should be categorized and ordered.

Modifying a Set

The settings used to create the set will be maintained for that set until those settings are revised. This means that the stacking, stamping, sorting, categorizing, and markup transferring behaviors will be the same for added sheets as they were for the initial sheets. If the user desires a different behavior, he or she can edit those settings using the Preferences icon ⚙ at the top of the Sets tab.

Clicking this button will open the Preferences window and automatically select the Sets window from the list on the left, as shown in Figure 5-14. There is a second related window titled Sets (Advanced) listed directly below Sets in the list. Between these two windows, all the options for the set may be adjusted and defaults may be set for the creation of future sets.

Whole Set Features

Finally, there are several features that allow the user to work with the whole set at once. Searching, selecting, and printing can be done on all the files at once without opening each individually. Additionally, there are features built in that allow the user to choose to include all the sheets, only the current sheets, or only the previous sheets. It can be very handy for publishing or printing the current working set of drawings, or archiving the previous drawings to free server space.

Combining and packaging are also available options. Combining simply combines all the individual PDF files into one file and packaging places all the individual files into a PDF envelope of sorts, similar to a zip folder. Both options allow the user to send the set of PDF files in a way that allows a user without Bluebeam to review and comment on the files. Currently, the Sets functionality is unique to Bluebeam and the .bex files cannot be utilized with other PDF software.

The last feature to highlight before moving on to discuss tags is the Drawing Log. As Revu manages revisions of drawings, it also keeps a tabular record of what it's doing, automatically

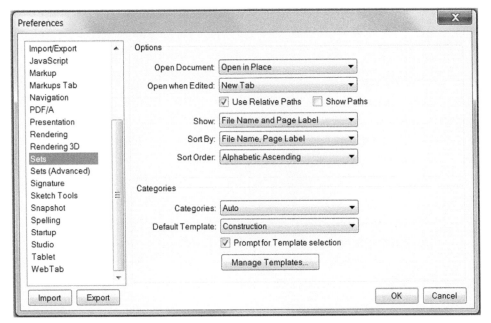

Figure 5-14: Set Preferences Window

listing all the sheets in a set, identifying their file name, page label, revision number, and any tags that have been associated with the file. Best of all, the log may be exported to PDF or Microsoft Excel and can include hyperlinks to each drawing. For Bluebeam users on iPad or Mac, or those without Bluebeam at all, this feature can provide great efficiencies for accessing drawings. The various settings in the Export Drawing Log window are shown in Figure 5-15. A sample Drawing Log in Excel format is shown in Figure 5-16. The reader may find further discussion of the Drawing Log in the "Tags" section of this chapter.

Tags

Tags are a wonderful feature that Bluebeam developers added to Sets with the Revu 2016 release. Tags are a unique way to store a piece of information about a PDF file by extracting that information directly from the file. Further, because tags are a subfeature of Sets, they operate in batch style, capturing the same information from each file in a set. The best way to understand this is to consider an example.

In the section "Creating a Set," we noted the Add Tags window shown in Figure 5-17 and suggested the user skip the tags or read on to the end of the chapter. That window is where this segment on tags will pick up. The first default Add Tags window will say, "Sheet Name" and "Select the page region that contains Sheet Name" and give the user the option to Select or Skip Page, Skip Document, or Skip Tag. In the majority of cases, the user should choose Select.

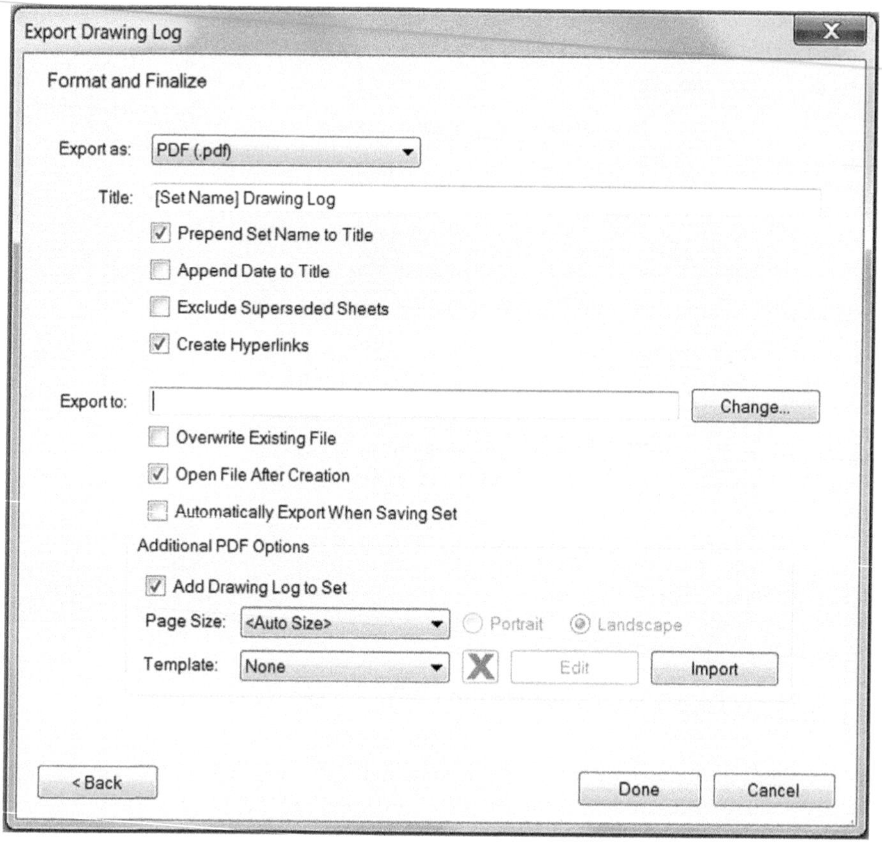

Figure 5-15: Export Drawing Log Options Window

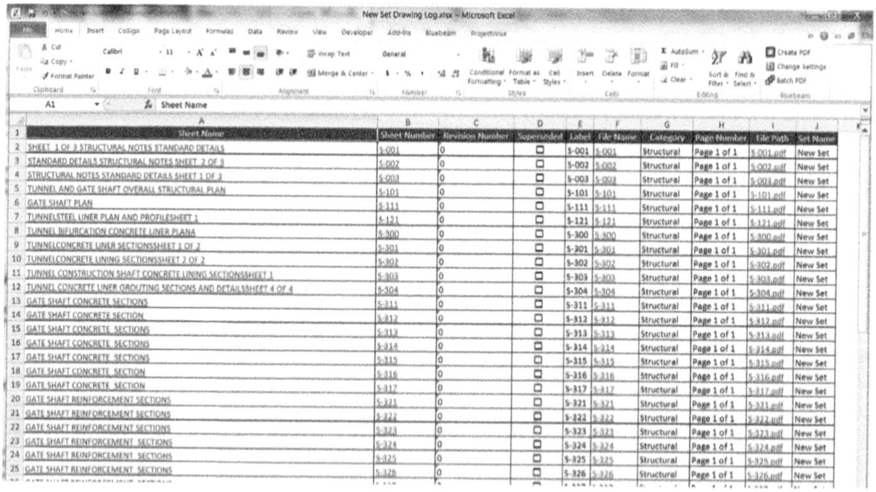

	Sheet Name	Sheet Number	Revision Number	Superseded	Label	File Name	Category	Page Number	File Path	Set Name
2	SHEET 1 OF 3 STRUCTURAL NOTES STANDARD DETAILS	S-001	0	☐	S-001	S-001	Structural	Page 1 of 1	S-001.pdf	New Set
3	STANDARD DETAILS STRUCTURAL NOTES SHEET 2 OF 3	S-002	0	☐	S-002	S-002	Structural	Page 1 of 1	S-002.pdf	New Set
4	STRUCTURAL NOTES STANDARD DETAILS SHEET 1 OF 3	S-003	0	☐	S-003	S-003	Structural	Page 1 of 1	S-003.pdf	New Set
5	TUNNEL AND GATE SHAFT OVERALL STRUCTURAL PLAN	S-101	0	☐	S-101	S-101	Structural	Page 1 of 1	S-101.pdf	New Set
6	GATE SHAFT PLAN	S-111	0	☐	S-111	S-111	Structural	Page 1 of 1	S-111.pdf	New Set
7	TUNNELSTEEL LINER PLAN AND PROFILESHEET 1	S-121	0	☐	S-121	S-121	Structural	Page 1 of 1	S-121.pdf	New Set
8	TUNNEL BIFURCATION CONCRETE LINER PLANA	S-300	0	☐	S-300	S-300	Structural	Page 1 of 1	S-300.pdf	New Set
9	TUNNELCONCRETE LINER SECTIONSSHEET 1 OF 2	S-301	0	☐	S-301	S-301	Structural	Page 1 of 1	S-301.pdf	New Set
10	TUNNELCONCRETE LINING SECTIONSSHEET 2 OF 2	S-302	0	☐	S-302	S-302	Structural	Page 1 of 1	S-302.pdf	New Set
11	TUNNEL CONSTRUCTION SHAFT CONCRETE LINING SECTIONSSHEET 1	S-303	0	☐	S-303	S-303	Structural	Page 1 of 1	S-303.pdf	New Set
12	TUNNEL CONCRETE LINER GROUTING SECTIONS AND DETAILSSHEET 4 OF 4	S-304	0	☐	S-304	S-304	Structural	Page 1 of 1	S-304.pdf	New Set
13	GATE SHAFT CONCRETE SECTIONS	S-311	0	☐	S-311	S-311	Structural	Page 1 of 1	S-311.pdf	New Set
14	GATE SHAFT CONCRETE SECTION	S-312	0	☐	S-312	S-312	Structural	Page 1 of 1	S-312.pdf	New Set
15	GATE SHAFT CONCRETE SECTIONS	S-313	0	☐	S-313	S-313	Structural	Page 1 of 1	S-313.pdf	New Set
16	GATE SHAFT CONCRETE SECTIONS	S-314	0	☐	S-314	S-314	Structural	Page 1 of 1	S-314.pdf	New Set
17	GATE SHAFT CONCRETE SECTIONS	S-315	0	☐	S-315	S-315	Structural	Page 1 of 1	S-315.pdf	New Set
18	GATE SHAFT CONCRETE SECTION	S-316	0	☐	S-316	S-316	Structural	Page 1 of 1	S-316.pdf	New Set
19	GATE SHAFT CONCRETE SECTION	S-317	0	☐	S-317	S-317	Structural	Page 1 of 1	S-317.pdf	New Set
20	GATE SHAFT REINFORCEMENT SECTIONS	S-321	0	☐	S-321	S-321	Structural	Page 1 of 1	S-321.pdf	New Set
21	GATE SHAFT REINFORCEMENT SECTIONS	S-322	0	☐	S-322	S-322	Structural	Page 1 of 1	S-322.pdf	New Set
22	GATE SHAFT REINFORCEMENT SECTIONS	S-323	0	☐	S-323	S-323	Structural	Page 1 of 1	S-323.pdf	New Set
23	GATE SHAFT REINFORCEMENT SECTIONS	S-324	0	☐	S-324	S-324	Structural	Page 1 of 1	S-324.pdf	New Set
24	GATE SHAFT REINFORCEMENT SECTIONS	S-325	0	☐	S-325	S-325	Structural	Page 1 of 1	S-325.pdf	New Set
25	GATE SHAFT REINFORCEMENT SECTIONS	S-326	0	☐	S-326	S-326	Structural	Page 1 of 1	S-326.pdf	New Set

Figure 5-16: Sample Drawing Log

Figure 5-17: Add Tags Window

Choosing Select returns the user to the PDF page and reverts the cursor icon to a cross-hair selection tool. As the dialogue box instructed, the user should draw a box around the sheet name on the page. The user should note that the zoom and pan functions will operate exactly as normal, so he or she should adjust the view as needed to conveniently select the Sheet Name, as shown in Figure 5-18.

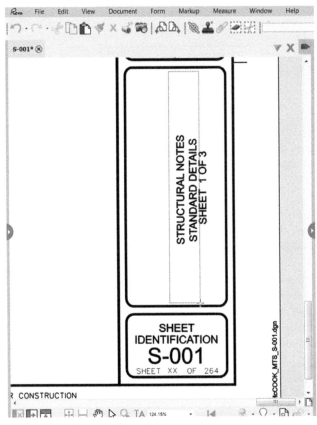

Figure 5-18: Selection of Sheet Name Tag

Releasing the mouse button will launch a second Add Tags dialogue box that will contain a Preview of the Sheet Name extraction. In the example, the sheet name was "STRUCTURAL NOTES STANDARD DETAILS" and, as shown in Figure 5-19, the extraction correctly captured the information. If the capture is correct, the user should choose OK to continue with the remaining tags. If the capture is incorrect, Revu provides the option to Reselect the Sheet Name.

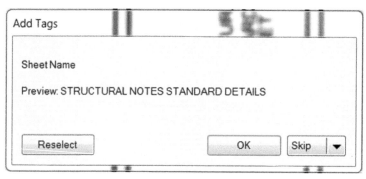

Figure 5-19: Add Tags Sheet Name Preview

EXPERT TIP

Selecting a Tag

Although remarkably simple once a user builds some experience, selecting a tag can be more of an art than a science.

A cool feature of the Tag extraction tool is its capability of reading vertical text. As shown in the example, text that is turned 90 degrees to fit in a sidebar title block on the left or right is not a problem for Revu. As such, the user shouldn't be concerned about the location of the title block.

As the user selects the first tag, the instinct is almost always to create a box that fits tightly around the title. That's generally great for the first page, but the user needs to acknowledge that the Sheet Name Tag, and all other tags, will be extracted from all pages of a set, not just the first one. The rectangular selection window will be used on every page at the same location in the same orientation. Therefore, the user needs to create the selection rectangle large enough to encompass the longest sheet name in the set. If the user had stacked the hard copy set on a light table, he or she would need to draw the rectangle on top such that all of the sheet titles in that set fit inside the projection of that rectangle. If the user discovers a tag that contains only part of the title, it's more than likely a matter of correctly sizing the selection box. See Figure 5-20 for an example.

To aid in the capturing of the largest sheet name, Revu does allow the user to page through the sheets to find the largest name and use that sheet for the selection box. To do this, simply click Select or Reselect and use the page forward/backward arrows at the bottom of the main window to progress from page to page.

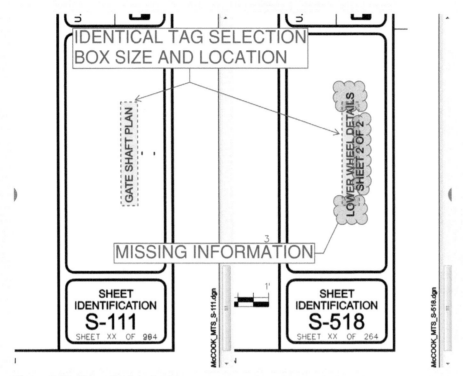

Figure 5-20: Sizing a Tag Selection Box

Once the selection window is correctly placed, the user may click OK. Revu will process that tag and move onto the next. Depending on the format of the files being used, the processing may be very brief or very quick. This is because OCR or optical character recognition is built right in. In some cases, Revu needs to convert raster text into searchable text in order to complete the extraction. In other cases, the text was created searchable and no conversion is needed.

In the event that Revu discovers a page that does not have text inside the selection rectangle, a second Add Tags window will appear with the troubled sheet displayed in the background. This may happen when the set of interest intentionally contains more than one sheet size, such as a mixture of Arch D (24 × 36) and Arch E (36 × 48) sheet sizes. A selection rectangle on Arch E would likely land off the page for size Arch D. It might also happen when the sheets are unintentionally printed with the wrong settings, perhaps centered differently or printed at the wrong scale incidentally, in either case resulting in an empty selection box on at least one of the sheets.

When this occurs, Revu assumes the title is in a different place and gives the user the opportunity to select a second Sheet Name rectangle for that particular sheet. Once the user selects OK again, Revu continues processing the current tag using the first selection rectangle as the primary source, and deferring to the second selection box when the first encounters a blank.

In some rare instances, the tag will not be applicable to a particular sheet. Maybe the cover sheet does not have an official Sheet Name shown on the title block. For these cases, the Skip options are very useful. Skipping can be applied to an individual page, a multi-page document, or the whole set. Skipping must be done one tag at a time. As such, skipping tags completely requires the user to skip once for each tag.

When skipping a page, Revu will simply move on to the next page and prompt the user to select the applicable page region on the next page. When skipping the document, Revu will move past all the pages in the current document, open the first page of the next document, and prompt the user again. When skipping the entire tag, Revu will move to the second tag and prompt the user for that selection box.

By default, the second tag that prompts the user is the Sheet Number. Following the same method as before, the user can define a rectangle that will encompass the sheet number on every sheet in the set and select OK to assign the correct tags to each page.

Revu will proceed through all defined tags, default or user-added, until it has captured each tag for each page in a set unless that tag or page has been skipped. Once complete, Bluebeam will prompt the user with the Tags window, a table of all the tags captured for all the pages within a set. A sample Tags table is shown in Figure 5-21.

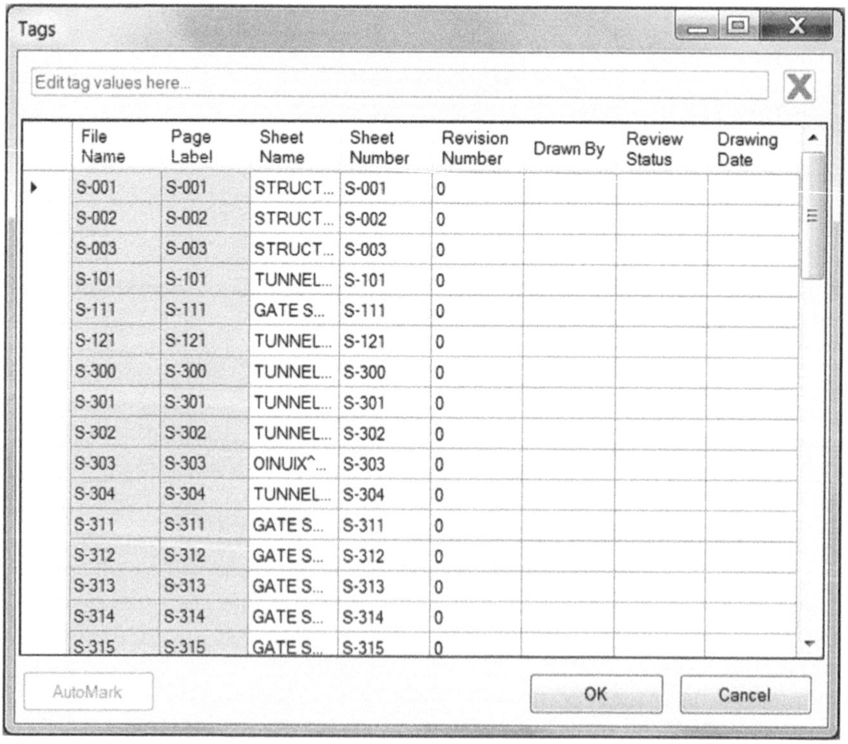

File Name	Page Label	Sheet Name	Sheet Number	Revision Number	Drawn By	Review Status	Drawing Date
S-001	S-001	STRUCT...	S-001	0			
S-002	S-002	STRUCT...	S-002	0			
S-003	S-003	STRUCT...	S-003	0			
S-101	S-101	TUNNEL...	S-101	0			
S-111	S-111	GATE S...	S-111	0			
S-121	S-121	TUNNEL...	S-121	0			
S-300	S-300	TUNNEL...	S-300	0			
S-301	S-301	TUNNEL...	S-301	0			
S-302	S-302	TUNNEL...	S-302	0			
S-303	S-303	OINUIX^...	S-303	0			
S-304	S-304	TUNNEL...	S-304	0			
S-311	S-311	GATE S...	S-311	0			
S-312	S-312	GATE S...	S-312	0			
S-313	S-313	GATE S...	S-313	0			
S-314	S-314	GATE S...	S-314	0			
S-315	S-315	GATE S...	S-315	0			

Figure 5-21: Tag Table Window

In the table, the user will always see the File Name and Page Label columns that are populated from the file metadata, not from data extraction. The adjacent columns will be the tags that were established, with extracted values in each cell. Here the user can manually edit any tags that were erroneous simply by clicking on the cell and typing in the correct tag value in the bar at the top. The user is also able to multi-select and multi-edit tags by selecting more than one cell with the Control or Shift key. Tags may be edited at any time using the Edit Tags button ⟍, in the top menu bar of the Sets tab, as shown in Figure 5-22.

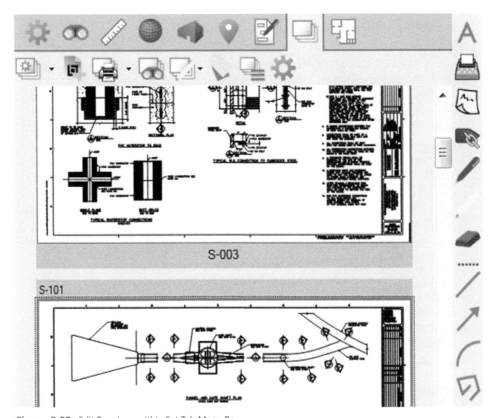

Figure 5-22: Edit Tags Icon within Set Tab Menu Bar

Finally, there will always be a Revision Number column that Revu will automatically populate. With the addition of tags, Revu is able to more easily track drawing and document revisions. If a page is added to the set, and its sheet number has not previously been part of the set, Revu will assume that the page is Revision 0. If a page is added to the set, and its sheet number has existed previously, Revu will determine the highest Revision number for that particular sheet number, and assign the subsequent number to the new sheet. For example, if S-501 has already been through the revision process twice, it's likely that there are three versions of S-501, Revision 0 the original, Revision 1 the first modified sheet, and Revision 2 the second modified sheet. The new version of S-501 would then be Revision 3.

It should also be noted to the user that as new sheets are added to a set, the tags are extracted automatically. Revu will remember the location of the Tag selection boxes, utilize the same boxes, assign the tags, and provide an updated Tag summary table.

Troubleshooting Tags

On occasion, tag extraction can be frustrating. As daily users of Bluebeam, the authors have encountered a few pitfalls that can be easily resolved.

The first pitfall is the garbage tag. On occasion a tag extraction will not return the expected text, but instead return a random sequence of symbols looking like binary code, wingdings, or an ancient computer language. In nearly every case the authors have encountered, this error is a matter of font type. In order for the tag extraction to work correctly, the font must be recognized by Bluebeam. A simple fix is to change the font used to create the PDF file. If this isn't possible, a potential resolution is to print the PDF, rasterizing the text and forcing Bluebeam's OCR engine to convert to searchable font once again. In some instances, the OCR engine has been able to capture a recognized font and in other cases, the OCR engine has captured the same garbage tag as before.

The second pitfall is the double tag. In a few occasions, a sheet number or name has appeared twice due to stacked text in the native file. When printed or converted to PDF, these text boxes overlap and look normal, but Bluebeam may recognize them as two and include the sheet number or name twice. Two simple fixes are to remove the extra text box in the native file or manually edit the tag in Revu.

The third pitfall is the slight shift. As mentioned previously in this chapter, it's highly important that each revision of each sheet is printed with identical settings as all the other sheets. A minor shift in the title block may be invisible to the human eye, but it may be enough to cause the Bluebeam tag selection window to miss part of the text, which may result in missing the entire tag for a given sheet. A simple way to check for sheet consistency is to zoom in on one corner of the title block and page through the drawings with the page arrows. This will jump from page to page to page and clearly indicate minor shifts in the title block or text within the title block. The simple fix is to create the drawings consistently.

The potential for tags is nearly unlimited. The user can establish tags that represent any piece of data shown on a sheet, from designer initials to client name; if it's on the sheet, it can be captured with a tag. Creative users might even begin to add data that hadn't existed previously or add data that isn't part of the sheet, by skipping the extraction step and typing it into the Tags summary table.

One use case the authors can imagine for the manual tagging is to track the review process of the drawings. For example, consider a tag named Review Status. When a sheet is originally added to the set (Revision 0), it has no Review Status. The sheet or drawing gets reviewed, comments are added, and when the review is finished, the reviewer manually changes the Review Status to "Reviewed." The updated status indicates to the designer that the comments are ready for incorporation. The designer updates the sheet based on the comments, adds a new revision (Revision 1) to the set, and changes the Review Status for Revision 0 to "Incorporated." The incorporated status indicates to the reviewer that the incorporated comments are ready for backcheck. The reviewer verifies the comments were incorporated correctly, adds any missed, incorrect, or additional comments to Revision 1, and changes the Review Status for Revision 0 to "Backchecked." At this point the Revision 0 sheet is complete and would only be used to look back for reference. Revision 1 moves on in the design process with the Review Status being updated each step along the way. The workflow is demonstrated in Figure 5-23.

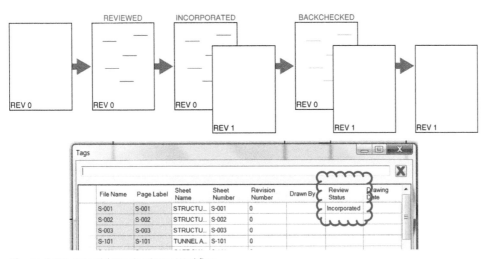

Figure 5-23: Potential Drawing Status Workflow

A bonus feature that came with the addition of tags is the ability to export a Drawing Log as discussed in the Sets section of this chapter. Essentially, this Drawing Log is a Tag summary that happens to work remarkably well as a drawing log for the sheets within a set. Further, almost any information desired for a drawing log can be captured in the Tag summary because all of the relevant information is captured on the sheets and may be extracted with a tag. Things like Drawn By, Checked By, Issue Date, Sheet Size, Client Name, File Name, and more, can all easily be captured in a nice, concise, easy-to-export table.

Meet Tyler Menard, Project Engineer, Sundt Construction, Inc.

Tyler Menard is a project engineer with Sundt Construction, Inc. in San Diego, California. Most of his days are spent on construction sites, managing projects, coordinating sub-contractors, and generally making sure progress stays on schedule.

Project Engineer, Tyler Menard

A big part of Tyler's job is keeping the construction documents organized—updating revisions, informing subcontractors, processing RFIs (request for information), and making sure everyone has access to the information they need. Before the introduction of tags, accomplishing that was somewhat of a computer syntax nightmare. "My time shouldn't be spent trying to name files exactly how they should be," said Tyler. In 2016, tags changed all that.

Tyler first heard of Bluebeam in college at Cal Poly, noting, "They introduced Bluebeam, but they weren't teaching any of the document management stuff." In fact, it wasn't until he started at Sundt that he learned Revu was capable of more than just markups.

"One of my colleagues told me to use sets. He showed me how to do it and said it worked well on his most recent project." Tyler discovered the same.

In a reality where a drawing set might be hundreds or even a thousand pages long, rendering the sheets takes time, sometimes up to a minute in the worst cases. Utilizing Bluebeam Sets, Tyler's subcontractors were able to render the sheets almost instantly, saving precious time and frustration. Revu's algorithm to open one sheet at a time made life easier for everyone involved.

Revu's Sets wasn't the only feature saving Tyler time; he also began using Bluebeam Studio Projects to share construction documents. "The old way was a large TV in the job trailer," said Tyler. "Subcontractors would come access the electronic files in the Sundt trailer, or ask for a memory stick with the drawings each time they were updated. They wouldn't be able to directly access the current drawings from their trailers." With Studio, Tyler's subcontractors check in on updates from their homes.

"It took some effort, but most of the subcontractors really liked the new setup," said Tyler.

When it was time to make the change, Tyler created a short PDF training document with instructions to download Bluebeam Vu, to access the construction drawing Set, and to access RFIs and ASIs (Architects Supplemental Information). He also hyperlinked the drawings and documents using Bluebeam's Batch Link and the manual hyperlinking tools. Some subcontractors learned the system faster than others, but will a little help, most of them came to understand and appreciate how it worked.

Tyler proceeded to use Sets and Studio on all of his subsequent projects. He limits all of the subcontractors to read-only access and in fact all of the other Sundt professionals too. Tyler said, "I wanted to protect the integrity of the Set. I don't want anyone deleting anything by accident. When a sheet gets deleted by accident, people think they have the latest, but they don't. That's when you have a real problem."

Tyler doesn't consider the read-only limitation a hardship at all, noting the ability to easily download a copy for markup at any time. In Tyler's mind, there was only one real problem and that was how much time it was taking him to manage the set. Each time revised drawings were issued, Tyler had to correctly name the files, import them into the set, and verify that they made their way to the correct location. "One extra space in the title would prevent the sheets from matching. I had to check every single update to make sure they were stacked correctly. It was frustrating and it was taking too much of my time."

When Bluebeam introduced tags in 2016, Tyler's frustrations disappeared. The file name was no longer important for matching versions of a drawing. The tags function would match the drawings automatically using nothing more than the sheet number in the corner of the title block.

Tyler's set management time was cut in half. His subcontractors got information faster. Construction documents were easier to navigate. Rework was reduced. It seemed to be better for everyone.

Tyler discovered unexpected benefits, too. "Drawing logs have been very helpful," he added. He can print out the log weekly and send it to the subcontractors as supplemental confirmation of which versions of the construction documents are the latest and greatest. The log has dates, sheet numbers, revision numbers, sheet names, and easily informs the contractor which sheets are current. "At that point," said Tyler, "it's the subcontractor's responsibility to make sure their drawings match the current set."

Even with all the benefits to date, he's not convinced he's done and he's still exploring what's possible. The next item in his plan is to start using Revu's drawing log as the sheet index because the hyperlinks are created automatically, saving him from doing that manually with every single update. He's also excited for custom tags and the ability to run reports on data that's in the actual drawings.

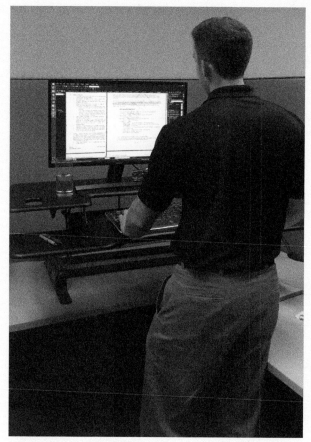

Project Engineer Tyler Menard

Tyler acknowledged that Bluebeam Sets has many benefits, but tags is what really enabled the workflows, saying, "Sets was great. Tags was an added bonus."

In addition to the features highlighted in this chapter, Tyler has found time savings from several of Bluebeam's other capabilities. The document comparison capabilities came in handy on Tyler's newest job, where the drawings were being updated without narratives or clouding. Without identifying all the changes, Sundt ran the risk of losing money or losing the project by bidding incorrectly. With the batch ability of the comparison tools, Tyler was able to quickly and easily identify every last change, giving him confidence that Sundt's bid was correct and up to date.

Tyler has also started providing RFIs and ASIs as attachments to the PDF files within a set. It keeps everything in one place and makes them easy for subcontractors to find and access. He flattens everything in the file except for the attachments, resulting in a markup list that serves as an RFI log. "It does increase the file size. I end up with single sheets that are as large as 50 MB because of attached RFIs, but it makes them so easy to find. There's no searching; it's a simple click."

Tyler has found that using Bluebeam gives him more time to do what he loves to do. "My mentor, Spencer Draper, always tells me, 'As engineers, we need to get out in the field and not sit at a computer all day managing documents.' Any way I can save time in the office and get out in the field is a huge benefit to me."

Conclusion

Sets, in conjunction with Digital Slip Sheeting and Tags, have changed the way companies execute work. They have simplified the creation and editing of a working drawing package, enabled the electronic working set, simplified the arduous process of tracking and managing changes, and streamlined the review process.

Readers who haven't explored sets should dig into the software and figure out how the feature can work for them.

Chapter 6
Issuing

In the architecture, engineering, and construction world, getting the right markups on a document is only half the process. When a document reaches the end of its revision cycle, the document owner must properly indicate the status of the document, possibly sign it, and send it to the correct location for record. Sometimes, the owner must send a summary of document changes to the client or the entire group. This chapter will explore the tools within Bluebeam that help approve and package up documents on their way out the door.

Stamps

AEC documents, especially drawings, use stamps extensively to indicate status, such as "Preliminary," "For Review," "Approved," or "Issued for Construction," as well as to mark a document as Confidential or Proprietary, and registered professionals must apply their seal to certain documents. Bluebeam provides many such stamps as a default collection that comes with the software. Check out the available default stamps under the Markup tab, Stamp button (Figure 6-1).

To apply one of these stamps, click on it from the Stamp list and then click on the document to place it. Some of the stamps are dynamic, meaning they have data such as date, time, and user that will automatically populate information at the time the stamp is place. Dynamic stamps do not update in real time; after the stamp is placed, the data becomes static.

If none of the default stamps are exactly what is needed, a custom stamp can be created. To do this, click on Create Stamp under the Stamp button. Fill out the information in the Create Stamp dialog box. The Template field allows a user to start from a common stamp format, like Text with Border. The dimensions of the stamp can be set here. A stamp can be resized after it is placed on a document but will keep its width to height ratio. Opacity and Rotation are helpful for stamps like "Draft," where the stamp is more of a watermark over the document. Click OK after setting up the Create Stamp dialog. A new PDF will open with the stamp area drawn as a box (Figure 6-2).

The stamp can be resized here as well by dragging the blue corners of the stamp area. Use the regular markup tools to create the stamp instead the stamp area. If a text box is drawn, the option will appear to add Dynamic text (Figure 6-3).

There are many types of text that can be populated dynamically at the time of stamp placement. Some of the most commonly used ones in AEC are User, Date, and Time, indicating who is stamping the document and at what date and time. The User field will populate as whatever the user's name is set as in the General tab of Preferences (Settings > Preferences > General). The syntax of dynamic text is an ampersand with the data field inside brackets (Figure 6-4).

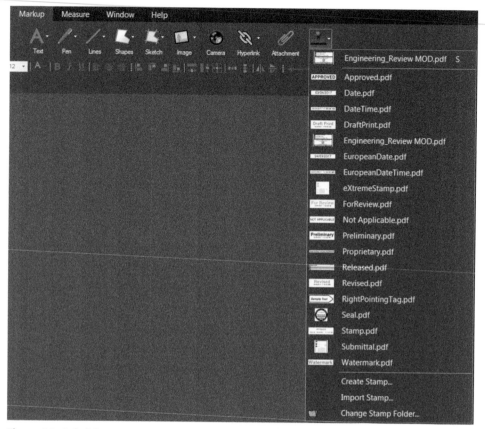

Figure 6-1: Default Stamps

Figure 6-2: New Stamp

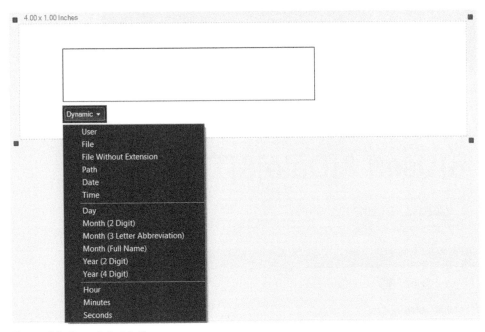

Figure 6-3: Dynamic Text Options

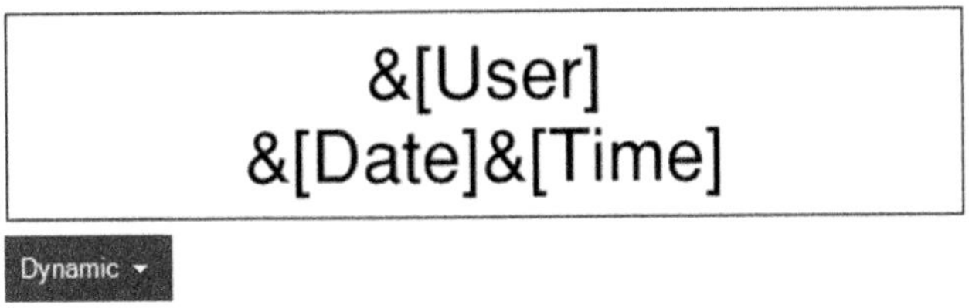

&[User]
&[Date]&[Time]

Figure 6-4: Dynamic Text Syntax

Images like company logos or photographs can be added to the stamp. When adding a text box to a stamp, it's a good idea to set the font size to Auto so the text will resize along with the stamp and allow for text that will vary in length, like dynamically populated user name. This can be done by checking the Auto box under the Properties tab when the text box is selected (Figure 6-5).

When the custom stamp is set up as desired, click Save. The stamp will now appear ready for use under the Stamp button.

A very useful custom stamp to create is one's professional seal. Electronically sealing documents is widely used and acceptable, depending on the client's or owner's preferences and PE Board bylaws (in the United States). To create one's electronic seal, the professional will need an electronic vector image of the seal. This can be created in AutoCAD, Visio, Microstation, and various other drawing software. After creating the seal in one of these programs, print

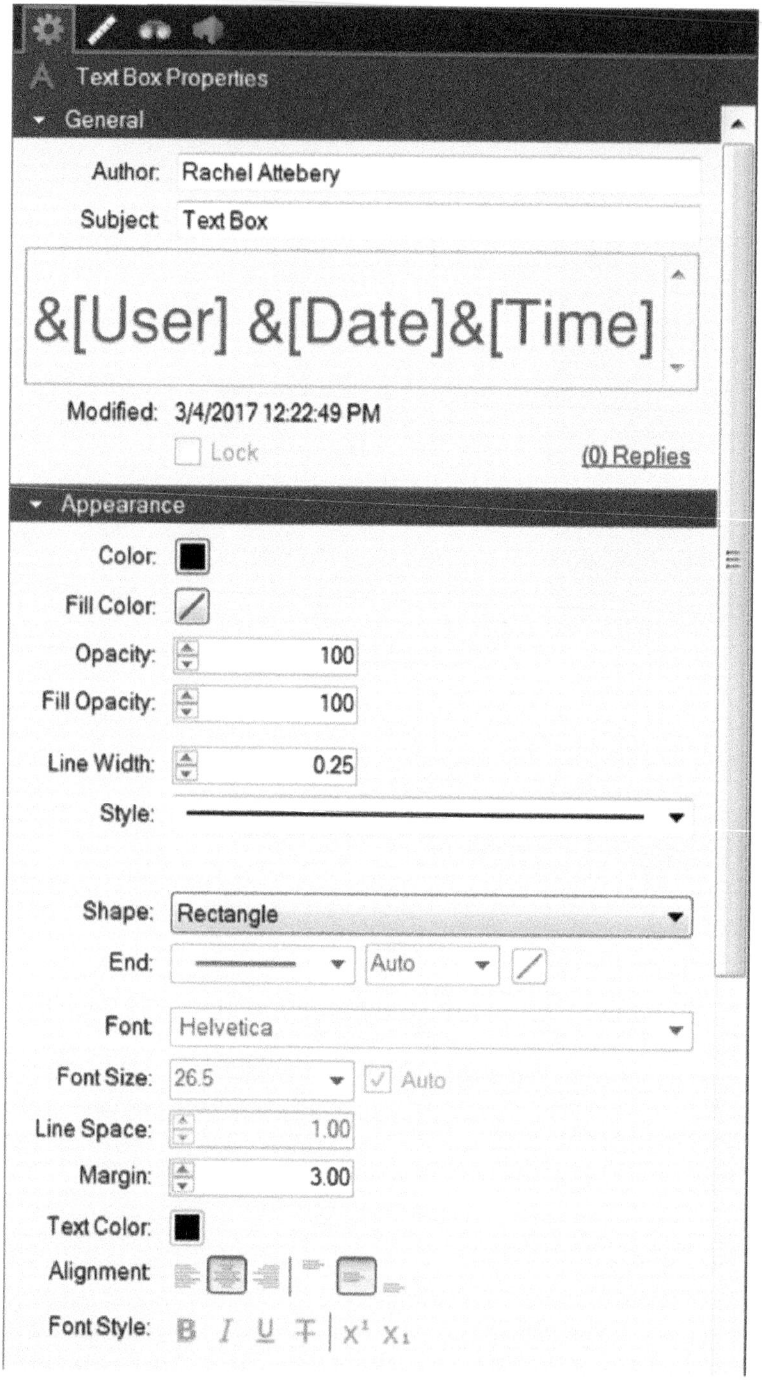

Figure 6-5: Auto Size Text

it to PDF. Crop the PDF down to the edges of the seal to create a square page shape with minimal white space around the seal. Save the PDF and then import it to the Stamp list. Click on the Edit button next to the stamp name (Figure 6-6).

Figure 6-6: Edit Stamp

This will open the stamp and allow the user to add a dynamic data field on top of the seal. Draw a text box where the date should appear on top of the seal. The Dynamic drop-down box will appear. Select a date option from the Dynamic menu. The date format can be customized by combining the options for Day, Month (2 Digit), and so on in the Dynamic menu. Add formatting as needed between the date pieces. For example, to create the date format "01/Mar/2017," add in the dynamic pieces Day, Month (3 Letter Abbreviation), and Year (4 Digit) and place the forward slash character "/" between each piece (Figure 6-7).

Figure 6-7: Dynamic Date over Seal

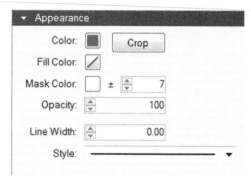

Figure 6-8: Mask Color Tolerance

When the dynamic date field is set up as desired, save the stamp. Depending on the professional's company policies, client/owner preferences, and local regulations, the professional can sign on top of his or her seal electronically or with a traditional wet signature. If it is necessary to use a wet signature, leave the seal as is and sign it after printing to paper. If it is acceptable to use an electronic signature, add the digitized signature image on top of the seal. A digitized signature can be captured with various hardware/software packages that allow the user to sign manually on an image capture field, which creates an electronic vector image of the signature. To add the signature image to the seal, click on the Image button under the Markup tab, navigate to the signature image, and position it on top of the stamp. Make sure the background of the signature image is transparent. If it is not, mask the white background by selecting the signature image and choosing white as the Mask Color in the Properties tab. The tolerance may have to be increased to fully remove the white background (Figure 6-8).

The signature image can also be color processed to make it blue or red or any color instead of black, which is likely the default color created by the signature capture tool. Do this before importing the image to the stamp.

Any PDF can also be used as a stamp. To add an existing PDF as a stamp, click Import Stamp at the bottom of the Stamp options. Navigate to the PDF, click on it, and click Open to add it to the stamp list. This is useful if one person creates a custom stamp and wants to share it with the team. The stamp creator can email or post the stamp in a shared location, and each project team member can import that stamp to their list. All stamps are stored in a folder, which is set up by default by Bluebeam Revu. This folder location can be changed by clicking on Change Stamp Folder at the bottom of the Stamp options. Navigate to the folder where the stamps should be stored and click Select Folder. The current stamp folder location can be discovered by copying and pasting the file path from the top of the Select Folder dialog box. Then the current stamp folder can be opened via Windows Explorer, and the user can manage his or her stamps inside the folder. Sometimes, the same stamp needs to be applied to many pages in a large multi-page document. This can be done quickly by right-clicking the stamp applied on the first page and selecting Apply to All Pages. The stamp will be duplicated on every page of the document in the exact same location. Notice that when a dynamic stamp is duplicated, the dynamic data does not change from the first stamp; the data will be the same as whatever was captured at the time of placement of the first stamp.

Flatten

After stamps are placed, it can be useful to Flatten the document before issuing it. Flattening makes markups part of the page, so they can't be edited or moved or even clicked on. Flattening is good for preventing recipients of the document from changing any existing markups. To flatten a document, go to the Document tab > Flatten button. The Flatten Markups dialog box will appear. The first, most important option to consider is the Allow Markup Recovery (Unflatten) checkbox. If this box is checked, anyone will be able to unflatten the document.

Consider the recipients of the document—are they likely to purposely unflatten the document and modify the markups if they know this is an issued revision? If yes, it's probably a good idea to uncheck the box and prevent markup recovery. It's also always possible to save a copy of the file, flatten it without the option for recovery, and send it out while retaining the unflattened copy to continue working from if necessary. Assign Layer allows the markups to be attached to any existing layers, or add a new layer from this dialog window. Even if the document is flattened, the layers will persist, and can be turned on and off to show markups assigned to those layers. Show Properties in Popup will retain markup properties, like author and date information, which will appear when the markup is clicked on, even though it cannot be edited. Unchecking this box will cause all markup information to be erased. Flatten Capture Media as Attachment allows the option to append any capture media to the file with hyperlinks to the media; that way, although the markups are flattened, the original media files can still be viewed. The user can choose which markups to flatten. Options include All Markups, skipping over any Filtered Markups, only flattening currently Selected Markups, and choosing flattened markups by type (image, stamp, text box, and so on). The user also has the option to flatten only certain pages of the document. Clicking Batch Page Range will open the Batch Flatten Markups dialog box and allow multiple documents to be flattened at once. For more help with the flatten options, visit http://support.bluebeam.com/online-help/revu2016–1/Content/RevuHelp/04—Document/08—Flatten/Flatten-Markups—MT.htm. Once all options are set, click Flatten. Open up the markup log and notice that the flattened markups no longer are listed there, because they are now considered part of the document and not markups floating on top of it.

Digital Signatures

Signing documents for approval is a big part of issuing. With the improvement in electronic document management technology and the Electronic Signatures in Global and National Commerce Act passed in the United States in the year 2000, documents can maintain a completely digital life cycle, from creation through approval. In the United States, only two states, Hawaii and Vermont, do not recognize digital signatures as a legitimate form of approval, as of August 2016. To check the electronic signature laws for a certain state, review the policy posted on the state's PE board website. The use of digital signatures should be judged on a project-by-project basis, considering client/owner preferences, company policies, and local jurisdiction.

Before walking through how to create and apply a digital signature in Bluebeam, it's important to understand the different levels of security for electronic and digital signatures. There are three main categories of signatures applied in the digital realm: simple electronic signature, first-party digital signature, and third-party digital signature. The term "electronic signature" encompasses all three categories. The category with the lowest security is the simple electronic signature. This would be a mere signature image with no further credentials associated with it. Simple electronic signatures include text boxes with names or initials typed into them, images of handwritten signatures pasted onto the document, and scans of wet signatures. Any hard-copy document that has been wet signed and then scanned only has simple electronic signature status in the digital world! The category with medium security is the first-party digital signature. A signature becomes "digital" when it has digital credentials associated with it in addition to the mere image of the signature. These digital credentials only exist in the digital world; in other words, if a digitally signed document is printed to paper, the paper copy has zero credentials attached to it. It now has only an image of the signature and is equal to a printed simple electronic signature. A first-party digital signature's credentials are generated by the user's computer and are stored locally on that device. Both Bluebeam Revu and Adobe products can generate this type of signature. Sending a document signed with a first-party digital signature to a recipient requires the recipient to "trust" the sender, meaning the recipient must take a manual action to let his or her computer know that he or she trusts the signer to be the person he or she claims to be. The category with the highest level of security is the third-party digital signature. A third-party digital signature's credentials are generated by a third-party certificate authority, such as DocuSign, GlobalSign, DigiCert, and many more. The third-party certificate authority does the work of verifying a signer is who the person claims to be, for a fee. The recipient of a document signed with third-party credentials does not have to manually trust the sender, as the certificate authority will do the work of matching the signer's information with its database of registered signatories. Again, digital credentials only exist in the digital world. Printing a digitally signed document to paper leaves the credentials behind.

Bluebeam supports the use of all three categories of electronic signatures. Inserting a signature image equates to a simple electronic signature. Setting up a first-party or third-party digital signature can be done through the Document tab > Signatures button > Manage Digital IDs. A user can set up multiple digital IDs with which to sign documents. This is useful if third-party signatures are required for some types of documents, but only first-party signatures are required for other types of documents. In an ideal world, third-party digital signatures would be used for everything, since they have the highest level of security. However, the fact that a company has to pay for the certificate authority, which often sets a limit on the number of signatures per year, can make it desirable to throttle the use of third-party signatures.

To add a digital ID for a third-party signature, the professional will first need to work through his or her company's certificate authority manager to get a third-party certificate set up. This usually results in the certificate being saved as a PKCS file (extension .PFX) on the company's network. After this is done, the professional can add the digital ID in Bluebeam by clicking the green plus sign in the Manage Digital IDs dialog box. Use the first option, "Browse for existing Digital ID file," and navigate to the saved certificate file. To add a digital ID for a first-party signature, open the same Manage Digital IDs dialog box, click the green plus sign, and select either "Create Digital ID file" or "Create Digital ID in Windows Certificate Store" (Figure 6-9). The first option creates a digital ID as a PKCS file, which can be used with any operating system and requires its own password. The second option creates a digital ID that is only compatible with Windows operating systems, but it does allow users to use his or her Windows login username and password to apply the digital signature. Fill in the Identity and PKCS Options in the New Digital ID dialog box. Click OK to create the digital ID.

Now, the professional has signature credentials set up and can start signing documents. However, because digital signatures are a set of digital metadata, a document can be digitally signed without showing any visual indication on the document of the signature. While the document is still technically "signed," it is best to create a signature image to go along with the digital credentials to provide a visual indication of the signature. To do this, click on the

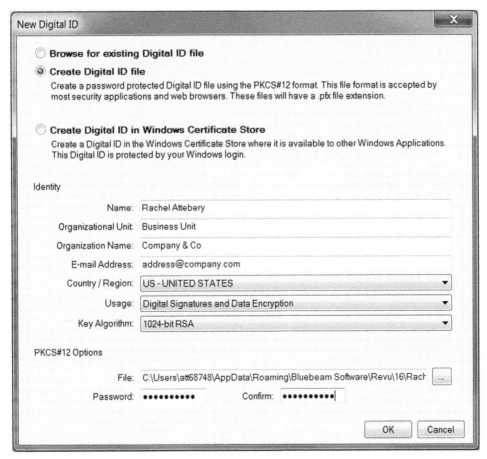

Figure 6-9: Creating a Digital ID

digital ID in the Manage Digital IDs dialog box. Click Manage Appearances. Click the green plus sign to add an appearance. Type in a name for this appearance, like "Drawing Approval," and modify the settings to create the desired appearance. A graphic can be included by uploading an image file, such as the professional's signature image. Various pieces of data like Name, Location, Reason, and so on can be included or not. Multiple appearance options can be set up for the same digital ID. This allows a professional to customize the look of his or her signature for different types of documents, while still using the same credentials in the background.

Once the appearance is set up, the user can apply the digital signature to a document. Remember, the point of a signature is to approve the content of a document at a certain time. Therefore, digitally signing a document prohibits modifications. If the document is modified after it is signed, the signature will be invalidated. Modifications include not only markups but also combining the document with other files, adding pages, deleting pages, adding bookmarks/page labels, adding additional signature fields, and anything else that would add to or subtract from the content that was approved at the time of signature. If multiple people need to sign the document, such as lead engineers from different disciplines, the first signer must set up all signature fields before signing the document. To do this, use the Add Signature Field function under the Signatures button and draw as many fields as needed. Add information to the fields by right-clicking on the field and selecting Properties. The Properties tab allows the form field to be named, have a tooltip appear when the signer hovers over the field, change if the field is visible or not, and make the field required (creates a red border around the field). After the signature fields are placed, the first signer can click inside the appropriate signature field. The Sign dialog box will appear. Select a digital ID and enter the password used to create the ID. Click Login. Add any additional info in the Options section. Select an appearance and click OK. When a document is digitally signed, it is saved as a new copy of that document. Select a save location. The document will now show the digital signature appearance, along with a green checkmark indicating that the signature is valid and verified (Figure 6-10).

Rachel Attebery
C=US, E=address@company.com,
O=Company & Co, OU=Business Unit,
CN=Rachel Attebery
Kansas City, KS
I am approving this document
2017.03.04 14:59:57-06'00'

Figure 6-10: Verified Digital Signature

If a user needs to clear his or her signature after signing, right-click and select Clear Signature.

When the signed document is sent to someone else, the recipient should validate the signature by right-clicking on it and selecting Validate Signature. If the signature is third-party, the signature should validate right away and display the green checkmark. If the signature is first-party, the recipient will need to add the signer's credentials to his or her trusted identities in order to validate the signature. To do this, the signer must export his or her certificate by clicking Export in the Manage Digital IDs dialog box. Save the certificate as a .CER file. Email or otherwise send the .CER file to the recipient. When the recipient receives the .CER file, he or she must save the file somewhere, then go to Trusted Identities under the Signatures button in Bluebeam Revu. Click the green plus sign to add a new trusted identity and navigate to the saved .CER file. Now the recipient can validate the sender's signature. Adobe has a similar validation process, and files signed in Bluebeam can be validated in Adobe and vice versa. In Bluebeam, all digital signature information applied to a document can be viewed in the Signatures tab (Figure 6-11).

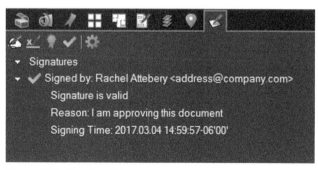

Figure 6-11: Signatures Tab

Assuming a digital signature is applied correctly and the document has not been modified after it was signed, the document will show a green checkmark over the signature field. If the signer's identity is unknown, or the signature has not been validated, or the document has been modified since it was signed, a question mark or yellow caution symbol will appear in the signature field. If the signature is invalid, a red X will appear. This symbol does not show up on printed paper copies of the file, but the signature image will. This is consistent with the fact that digital credentials only exist in the digital world. Considering a piece of paper to have valid digital credentials would be like expecting a physical paper clip sitting on the desk to help answer a question about Microsoft Word—it doesn't work that way. For more information about digital signatures, read Bluebeam's tutorial on digital signatures at www.bluebeam.com/us/bluebeam-university/pdf-tutorials/revu-12/digital-signatures.pdf.

EXPERT TIP

Digital Signature

A digital signature signs an entire file at once and does not need to be applied page by page. However, often a professional's signature needs to appear on each page of a multi-page document. When this is the case, it is best to first place any professional seals, dates, and signature images on the document on each page, then add a digital signature field and digitally sign the document with no signature appearance. This way, each page carries the traditional image of the seal, approval date, and professional's signature, while giving the entire document the credentials associated with the digital signature. For an even more efficient process, see chapter 10, "Go Digital, Engineering," for instructions on Batch Sign & Seal.

An alternative to digital signatures is certification. When signing a document, under Signature Type, instead of using Digital Signature, select Document Certification. The main difference between digital signatures and document certification is that document certification locks the content of the document but has options for what kind of modifications can be made to the document after it is certified. The most stringent option is "No changes allowed," which then essentially makes the certification equal to a digital signature, but also outright disables the ability to mark up the document. With digital signatures, the document can still be marked on after the signature is applied, but that action will invalidate the signature. With certification "No changes allowed," markups tools are completely disabled. The middle option is "Fill in forms and digital signatures," which allows subsequent reviewers the ability to fill in premade form fields, including digital signature fields, without invalidating the previous signatures. Markup tools are also grayed out with this middle option. The most lenient option is "Markups, fill in forms, and digital signatures," which still locks the original document content but allows additional markups, filling of premade forms, and digital signatures. To prevent additional markups but allow subsequent digital signatures, choose the middle option, "Fill in forms and digital signatures." Complete the Options and Appearance choices as normal, then click OK. The signature field will show an orange ribbon over it, indicating the document has been successfully certified (Figure 6-12).

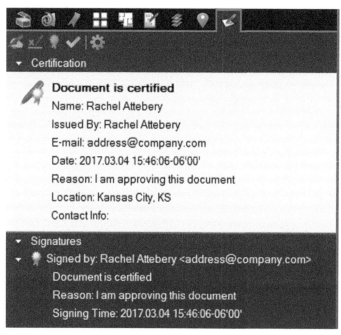

Rachel Attebery
C=US, E=address@company.com, O=Company &
Co, OU=Business Unit, CN=Rachel Attebery
Kansas City, KS
I am approving this document
2017.03.04 15:46:06-06'00'

Figure 6-12: Verified Certification

The certification also counts as a digital signature, as shown in the Signatures tab (Figure 6-13).

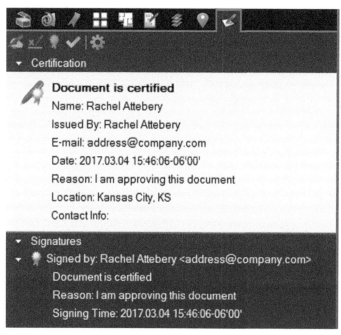

Figure 6-13: Certification in Signatures Tab

While a document can only be certified once, by the first user who applies his or her credentials, it can be digitally signed by subsequent signers, if signature fields have been set up by the first user. Certification is possibly the most foolproof method of digitally signing a document because it can remove the ability to mark up the document, which invalidates signatures and sends the document back to the beginning of the approval process.

Meet Jay Curebanas

Design Technologist, Black & Veatch

Photo Courtesy of Jay Curebanas

Jay's business unit at Black & Veatch decided to adopt a paperless work flow for telecommunications projects. This was fairly easy to implement, until the last step of the process: signing the document. Professionals could work paperlessly up to the point of signature, but had to print out the documents to wet sign them at the end of the process. Jay's department head identified the solution of digital signatures, and Jay focused on deploying Bluebeam Revu because of the automatic connection between Bluebeam Revu and major digital signature certificate authorities.

At the beginning of the deployment, the group didn't have a documented process or method for digitally signing documents. Because the documents are primarily engineering drawings, there is a higher level of complexity involved with sealing and signing. There are professional seal graphics and state requirements to consider and artwork format requirements to meet. Jay had to deliver all that and a digital signature at the same time, so he developed a process and trained the engineers. They went through a few evolutions of the process and ended up with Bluebeam stamps for the engineers' seals stored in personal Bluebeam Tool Chests for each engineer. This allowed the CAD technicians in Jay's group to eliminate the step of placing seals on CAD files for the engineers, and put the seals in the local machines of the engineers so they were protected and applied directly by the engineers. This created more security for the seal graphics. Jay notes that although forgery and malicious acts are tough to prevent against, this was a step in the right direction. It is very easy for someone to forge a hand-signed document, but with the new electronic method, a person with ill intent has to go through the extra steps to generate all the necessary seals and graphics.

As far as user adoption goes, it was clear that the engineers needed to be self-sufficient on the sealing and signing process. Jay wrote a guide to coach them on how to get their digital signature and graphics, and offered several methods to capture their signature image based on resources they might have available to them. To capture the signer's signature graphic, Jay recommends that signers get Bluebeam Revu for iPad, with a stylus, and draw their signature using the pen tool. This enhances the electronic workflow because the signature will be high-fidelity vector content rather than the raster content obtained by scanning in a wet signature. This is important when a signature graphic needs to be scaled up in size, to prevent the appearance of graininess or pixelation. Jay also created a Bluebeam Profile tailored for his business unit that has a custom tool bar with all necessary digital signature tools pinned right next to each other. The team had an immediate success rate of over 50 percent in adoption of the digital signature process.

In order to seal and sign a document, the engineer places the seal stamp, signature graphic, and date text box in the appropriate location on the document. The signer can apply these markups to the same spot on all pages of the document by right-clicking on the seal and selecting Apply to All. To apply the markups to only some pages, the signer can open the Thumbnails tab and use the Ctrl key to select the desired pages, then use Apply to All to place them only on the selected pages. Then the signer flattens the document to secure the seals to the document. This is an important step because without flattening, the seals remain editable, even after digitally signing the document. If the document is digitally signed, and then one of the seals is accidentally moved or edited, that invalidates the digital signature. Therefore, flattening is a critical step in the process both to prevent accidental invalidation and to protect the seal from being copied and pasted. After flattening, the signer draws the digital signature field on the front sheet of the document. At this point, all the necessary graphics—the seal, the signature image, and the date—are placed in the appropriate locations in the document, so the final step of applying the digital signature is only necessary to add digital encryption to the document and does not require any additional graphical appearance. So the signer applies a digital signature with a blank appearance. Jay says this isn't necessarily a recommendation, but it makes sense for his group's process because they already have all the necessary graphics in place on the document and only need the final digital encryption.

For multiple seals on the same document, the process is slightly different. Whereas in the paper process, the first engineer would apply his or her seal and wet sign each page, then pass it on to the next engineer to do the same, in the digital process, after the first digital signature is applied, any content added after that is considered a change and invalidates the first digital signature. The digital signature doesn't understand the concept of an additional seal as not being a change related to design—it's a change. What Jay's team does in this case is have all engineers apply their stamps and signature graphics and flatten them. At that point, all necessary graphics are in, and the final signer digitally signs the document to encrypt it.

Another challenge is having two separate files that need to be brought together and viewed as a single file. For example, there is an electrical sheet set that sits inside of sheets prepared by civil and structural engineers. The two digitally signed sheet sets cannot be combined post-signing, because the change invalidates all the signatures. Jay says, "You have to achieve a level of sophistication with that and you can do it effectively if you want to; it's just a bit more of a ballet."

Next, the signed document goes into the electronic document management system as a record. Sometimes the files do get printed. It's unavoidable in some situations. Some jurisdictions and municipalities require the printed documents for their review. Some clients request it. At least having the official copy as a digital record eliminates the need for Jay's team to store and save and provide physical storage and archival of these documents over time. Jay says with the new electronic process, you can hear a pin drop in the office. When he started working at Black & Veatch, he recalls hearing the Xerox machine run all day, but now the need for reprographics is becoming less and less.

Jay says, "Setting things up in a proper way helped with adoption and use. In some ways, you have to be a graphic designer to sign a construction document. You have to understand all these different tools and how they come together—how to turn a JPG translucent! You have to pass that knowledge to the engineers so they can actually use this thing." Jay notes that while he didn't face too much resistance in implementing the new digital process, some professionals were initially concerned about the legality of the digital signature. Leadership had to reassure them that the certificate authority and process was vetted and legal. Once professionals felt comfortable with that part, they could appreciate the fact that with the digital process, they were using the same program to do two things at once: both sealing and signing. Jay remembers that this efficiency gain made a lot of sense to the engineers. "Some of my biggest critics had been the older engineering crowd. But when you've spent a lifetime wet stamping and sealing, and then realize, 'Hey, you mean I can just do this?' They loved it! No one wants to get the hand cramp of signing 35 sheets over and over! These engineers were trying to save any possible time out of their day."

Today, through Jay's digital process, things get sealed and signed much faster. The final deliverable is a much higher quality. There used to be stacks of folders and papers on desks and that's how reviews were done. That's not true anymore—everything is digital. Working with engineers remotely is much faster and easier, since the electronic file can be dropped in an electronic document management system, opened by the remote engineer, and sealed and signed right there. Bluebeam Revu's 2017 digital signature tool makes the process even more robust, allowing users to get creative with the tools they have and get the job done faster and better.

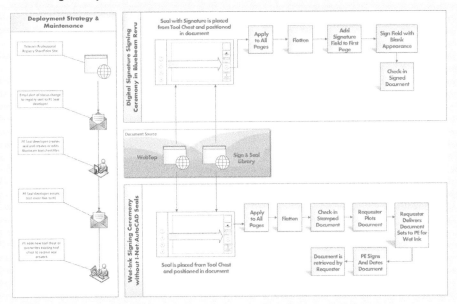

Jay's Digital Signature Process

Graphic Courtesy of Jay Curebanas

Document Management Systems

After a document is approved, it can be stored in an electronic repository of record. While that repository system varies from company to company, Bluebeam has chosen to integrate its product with two popular systems: Bentley's ProjectWise and Microsoft's SharePoint. The integration means that PDFs can be opened directly from these systems into Bluebeam, with the check in/out dialog controlled inside Bluebeam. A user can modify the document in Bluebeam, then check it directly back into the record system without having to save the file to the desktop first. To add a ProjectWise DataSource or SharePoint document library integration, go to Settings > Preferences > File Access (Figure 6-14).

Any existing integrations will be shown in the Document Management Systems window. The option "Toolbar integration only" determines how the user interacts with the integrated systems. If this box is unchecked, a dialog box will pop up offering the different system options (ProjectWise, SharePoint, Disk) every time the user opens or saves a file. If this box is checked, the user will use the Document Management toolbar to open, save, and check in files when needed. The option "Always show selection dialog" causes the system options to always pop up in addition to using the Document Management toolbar. "Enable batch check-in when closing Sets" is helpful when a user has modified multiple documents in a set and can check them all back in at once instead of going file by file.

To add a new system integration, click the Add button. Select either SharePoint or ProjectWise from the Type dropdown menu. If selecting ProjectWise, click Load to load

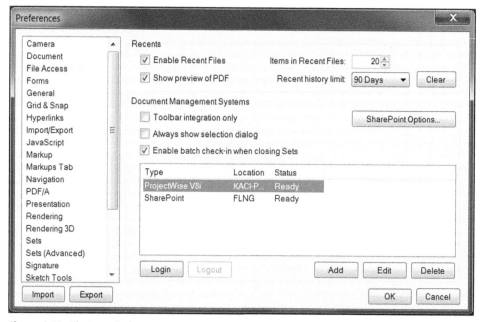

Figure 6-14: File Access Preferences

all available DataSources. The DataSources may appear with their technical names instead of the nickname that normally appears when working in ProjectWise. The first time a user opens Bluebeam, if that user also works with ProjectWise, Bluebeam will detect ProjectWise installed on the user's computer. A ProjectWise DataSource can also be added through this dialog box, if the user recognizes what to do when it appears. Otherwise, the DataSource can always be added later via the File Access Preferences. Type in the login name and password normally used to login to ProjectWise, or leave this blank to use Single Sign On. A couple of checkout options can be managed when adding the DataSource: Checkout on Open will automatically checkout PDFs whenever they are opened from ProjectWise into Bluebeam, and Checkout on Open from Hyperlinks will automatically checkout files that are stored in ProjectWise but are accessed via hyperlink. Click OK, and the new ProjectWise DataSource should appear in the Document Management Systems window. In order to work in ProjectWise from Bluebeam, click on the ProjectWise line item and click Login. A successful login will show "Logged In" under Status in the Document Management Systems window. To stop getting the option to work in the added system, click on it, log out, and delete it.

To add a SharePoint integration, choose SharePoint as the Type. Copy the SharePoint site URL and paste it into the Site field. Only include the URL address up to the point of the site name. Including the complete URL down to the .aspx extension will not work. For example, if the document library is found at URL https://aeco.sharepoint.com/groups/enterprise/ projectname/Project%20Document%20Library/Forms/AllItems.aspx, only include the URL https://aeco.sharepoint.com/groups/enterprise/projectname. Enter a nickname for the site as it will appear in Bluebeam. Choose a default library if desired to provide a shortcut to the most often used document library. Enter the login name and password normally used to login to this SharePoint site. Again, check the boxes by the checkout options if desired, and click OK. There are some additional options for SharePoint under the SharePoint Options button in the File Access Preference window (Figure 6-15).

If a professional notices that every time he or she works in SharePoint, the next time he or she works in Bluebeam, that SharePoint site has been added to his or her integrations, uncheck the box for "Auto detect SharePoint network paths."

If the Document Management toolbar isn't visible, unhide it by going to View tab > Toolbars button > click on Document Management. This will allow the user the ability to log in and out of integrated systems, open files from those systems, save files to those systems, and check in or out files from those systems without having to open the File Access Preference window. To work with files in ProjectWise or SharePoint, log in to the system from the Bluebeam Document Management toolbar. Click the Open File button to get an Open File from System dialog box. Navigate through the system folders, noticing that only PDFs can be opened. Multiple PDFs can be opened at once. When the PDF is opened in Bluebeam, notice the symbol in the Document tab, which will be a padlock or a red checkmark. The padlock indicates that the file is simply open, while the checkmark indicates it is checked out. Clicking on the symbol gives options to check out the file if it is not already checked out, check it back in, update the server copy without checking it back in, undo check out, look at the version history, and look at the file properties. Checking a file in and out via Bluebeam will still maintain the version history in its system of origin.

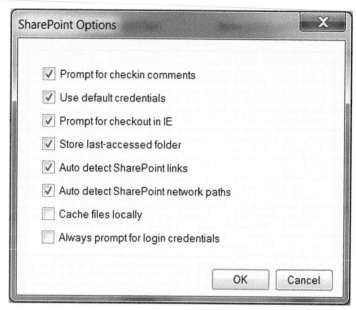

Figure 6-15: SharePoint Options

Any Bluebeam Studio Projects that the user is working with will also appear as options in the Document Management toolbar. In the same way that PDFs can be opened directly into Bluebeam from ProjectWise and SharePoint, they can also be directly opened from Studio Projects.

Summary

The document has now been stamped, signed, and uploaded to the correct repository. Sometimes that's not enough. Especially on large projects where many parties are involved, it is difficult to stay up to date on the status of all documents and be aware of every change that is made. To help manage change awareness, Bluebeam offers a Summary feature that automatically creates a list of all markups on a document and allows export of the list to CSV, XML, PDF, and directly to paper. Summaries can be run even after the document is digitally signed. To create a summary, open up the markup log of the document in the bottom flyout. Click on the Summary button (Figure 6-16).

Here is a rundown of the Summary options:

- CSV: Export the markup log as comma-separated values in rows and columns with no formatting.
- XML: Export the markup log as an .xml file, which retains formatting and can be imported to other programs as a table.

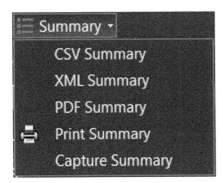

Figure 6-16: Summary Button

- PDF: Export the markup log as a PDF with options for table or flow layout and a snapshot of each markup.

- Print Summary: Export the markup log with the same options given by the PDF option, but print directly to paper instead of to PDF.

- Capture Summary: Export the markup log in PDF flow style with any captured images attached to markups shown as well.

For each summary type, there is the option to exclude filtered markups and hidden columns. This refers to how the markup log is set up before summarizing it. Each column in the markup log can be turned on or off by clicking on the Columns button and toggling on/off the various columns. Markups can also be filtered by turning on the Filter button and then clicking the blue arrow dropdown on a column header to filter by author, date, page number, and so on. These filter and hide options allow the markup log to be condensed in order to focus on certain information. This can be helpful if multiple audiences need to receive different types of information in their summaries. Sorting by Status can also be helpful in order to only show Approved marks or Rejected marks. Once the markup log is set up with the desired content, any and all of the summary options can be used.

One summary option to understand is Padding, which appears for PDF, Print, and Capture. Padding describes how much document context is shown around the snapshot of the markup. For example, on piping and instrumentation diagrams, having a zoomed-in snapshot of an added valve is of no value. The snapshot is only valuable when the valve is seen in the context of the larger system. Increasing the padding number will increase the document context; decreasing the padding number zooms in closer on the markup. Also in PDF and Print, there is an option to Include Capture Media Addendum. This is the same as running the Capture Summary and appending it to the PDF or Print Summary. The option in PDF and Capture to Attach Media as Linked Files will hyperlink the original photo or video files to the PDF Summary, which allows these files to be further used, or simply viewed in the case of video. If this option isn't checked, videos will be summarized with a still image. For all options except Capture, there is a box to Include Totals. This is used especially in estimating, when there is a grand total of counted items.

XML Summary

If XML Summary is used, the data can be imported into Excel formatted as a filterable table. To do this, open Excel and make sure the Developer tab is visible (for instructions on how to turn on the Developer tab, visit https://support.office.com/en-us/article/Show-the-Developer-tab-e1192344–5e56–4d45–931b-e5fd9bea2d45). Open the Developer tab, click Import, navigate to the saved .xml file, and click Open. For more information, visit https://support.office.com/en-us/article/Import-XML-data-6eca3906-d6c9–4f0d-b911-c736da817fa4#bmimport_an_xml_file_as_an_xml_list_wit.

Sometimes there are multiple individual documents that need to be summarized. This can be done in one fell swoop by using the Batch Summary function. Go to the File tab, Batch button, Summary. Add all the files to be summarized and click Next. The next window allows columns to be turned on and off and reordered. If there is a certain configuration of columns and data that has been saved for repeated use, click Load Config to navigate to that file and automatically set up the columns and their data. If not, click Next. The next window allows filtering of column data. The Sort dialog allows data to be sorted by multiple criteria, for example, first by file name, then by page label, then by author, and so on. Click Next. The next window allows selection of the export type: CSV, XML, PDF, or Print. Check the box next to "Create Multiple Reports per File Name" to create a separate summary file for each document. For PDF and Print, if there is a certain template to be used, click Import to import the PDF template. Fill out the other options as normal. If this summary configuration will be reused, click Save Config and save the configuration as a .bcf file. If this same set of files will often be summarized, click Save Batch so that all the files can be saved under a .bcx file and automatically uploaded to the Summary engine by using the Load Batch button at the beginning of the Batch Summary process.

Export

The last issuing feature is Export, which allows a PDF to be exported to a different file type: PDF/A, TIFF, JPEG, PNG, GIF, BMP, Text, RTF, HTML, Word Document, Excel Workbook, or PowerPoint Presentation. This is an incredibly powerful feature that allows fantastic versatility! PDF/A may be useful for exporting files for long-term electronic storage. An image file may be useful for working with customer-specific document management systems. Word and Excel are lifesavers if a user doesn't have the native file from which the PDF was created, and needs to edit the document beyond what can be done within the PDF. Word, Excel, and PowerPoint have the additional option to export the entire PDF or only a section of it. This is useful to export just a paragraph, or just a table from a PDF. Export options, such as image resolution, can be controlled in Preferences > Import/Export.

Conclusion

Bluebeam was created to allow engineers to retain a document's original quality by keeping it in the electronic world. Since its inception, additional tools and features have truly enabled an end-to-end digital process, from the creation of a drawing, through its review and approval and storage cycle. While each user and group and company will likely find their own custom process that best fits their unique use cases, the fundamental capabilities outlined in this chapter will support a more responsible way of working in the AEC industry.

Chapter 7
Measuring and Estimating

If electronically generated drawings are the basis for takeoffs and estimating, why print them to paper and use a ruler and manually count symbols to put together a material takeoffs list? Bluebeam Revu provides the tools to measure, count, and calculate costs for any piece of content on the drawing, while keeping everything electronic. Users can calibrate a drawing and then measure lengths, areas, perimeters, diameters, angles, radii, and volume, as well as count objects. The markups list's custom columns allow the user to assign monetary value to distinct types of materials and then provide subtotal and total cost estimates within the markups list for each material type. Revu 2017 introduced several new tools specifically for estimators, including Dynamic Fill for measuring complex regions, Quantity Link (eXtreme only) for connecting Bluebeam measurements to Excel, and enhanced measurement and count tools. For a recap of the new features, check out www.bluebeam.com/solutions/revu2017. This chapter was written using Revu eXtreme 2017 in order to give the most powerful and up-to-date picture of the measuring and estimating capabilities.

Calibration

The best place to start is with the Measurements tab in the right-hand flyout (in right-hand flyout by default, otherwise wherever the user has docked it). If the Measurements tab isn't showing, right-click in the blank space of the tabs toolbar, click Show and show the Measurements tab (Figure 7-1).

Before any accurate measurements can be taken, the user must calibrate the measurement tool. Do this by clicking the Calibrate button in the Scale section of the Measurement tab. Bluebeam will give instructions to select two points. This makes it easy if the drawing has an image of the scale. Notice that if it is vector content the cursor automatically snaps to the page content. The vector content aspect of measuring is very important, because it ensures the most accurate measurement and makes it easy to snap to the content. Whenever possible, use a drawing printed directly to PDF from the CAD program and avoid using scans of paper drawings. Select the starting and ending point for the calibration. The Calibration window will appear (Figure 7-2). Enter the at-scale measurement for the length selected. In this example, the 1-inch segment is equivalent to 10,000 mm in real life.

Also within the Calibration window, there is an option to Store Scale in Page. Almost always, the user will want to check this box, because it will save this calibration to the page, and next time the drawing is opened, the user can continue on measuring without having to recalibrate. The nice thing about storing the scale is that it only applies it to the current page, not all pages in a multi-page document, unless the user also checks the box for Apply Scale

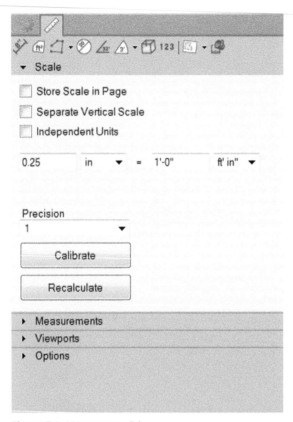

Figure 7-1: Measurements Tab

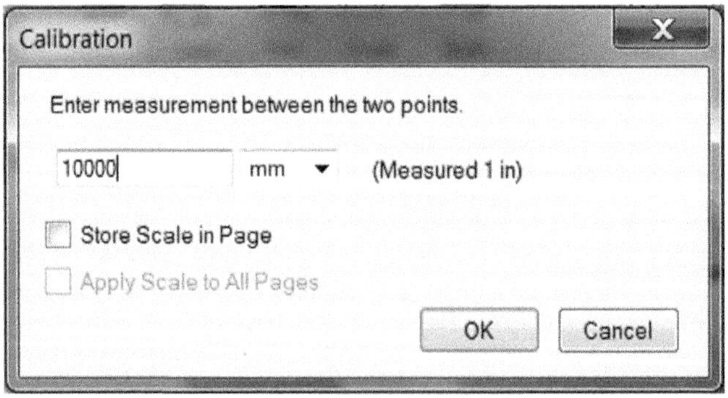

Calibration

Enter measurement between the two points.

10000   mm ▼   (Measured 1 in)

☐ Store Scale in Page

☐ Apply Scale to All Pages

OK   Cancel

Figure 7-2: Calibrating

to All Pages. Often, different pages in a large document will have different scales, but if that is not the case, Apply Scale to All Pages will save time for calibrating the other pages in the document. Click OK to finalize the calibration and start measuring. Do a test run to make sure the calibration worked correctly. Select the Length tool from the Measurement tab and

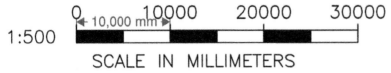

1:500

SCALE IN MILLIMETERS

Figure 7-3: Verified Measurement

select the same two points used in calibration. If the calibration was done correctly, the length will show the right real-life length for the measured segment (Figure 7-3). Hit the Esc key to exit the Length tool.

Measurement Tools

Properties

At this time, the user may notice that the color needs to be changed or the text of the measurement call-out is too big or in an inconvenient location. Change properties of the measurement text by clicking on the measurement, then opening the Properties tab in the right-hand flyout. Alternatively, the user can right-click on the measurement and click Properties. In the Appearance section of the Properties tab, change the color, opacity, line width, end cap style, font, font size and color, and so on as desired (Figure 7-4). The Highlight checkbox is useful, as this will retain the chosen color of the measurement, but make it transparent, keeping the drawing content visible below the measurement markup. Hold down the Shift key and use the mouse to move the text of the measurement markup without moving the entire markup. The entire text can actually be toggled on and off by checking or unchecking the Show Caption box. This can be useful for measuring at scale and totaling up quantities in the markups list while keeping the drawing itself clean and clear. The text can be dragged away from the measurement, again using the Shift key, and this will automatically create a leader line from the text back to the measurement. Turn off the leader line by unchecking the Show Caption Leader Line box. If using an end cap style, the size of the end cap can be changed by changing the line width, or by selecting a different option besides Auto in the dropdown menu next to the Start and End cap style boxes.

Figure 7-4: Measurement Properties

If the user plans to use this formatting on all other measurements of this type (i.e., Length measurements), he or she can scroll to the bottom of the Properties tab and click the Set as Default button. Now, whenever the Length tool is used again, it will retain the formatting set on this first measurement. What's more, the user can save this tool to his or her Tool Chest for quick reuse. Before doing this, the user may want to give this tool a specific name, like Trim, for example. The great thing about saving off tools like this to the Tool Chest is that it will help with totaling different types of material lengths later. Since the Length tool might be used to measure multiple types of trim or different types of walls, it's useful to save off the specific tools as they're created. To give this tool a specific name, change the Subject under the General section of the Properties tab. Then right-click on the measurement and click Add to Tool Chest and choose a Tool Chest to save it to. When the tool is added, it can go in as Drawing Mode or Properties Mode (Figure 7-5). Drawing Mode allows the user to duplicate the markup 100 percent, including the previous length. Properties Mode allows the user to use all the same properties but draw a new measurement with a new length. Properties Mode is usually better for measuring and estimating purposes. To change the mode, right-click on the tool in the Tool Chest and select Properties Mode (or Drawing Mode to switch back).

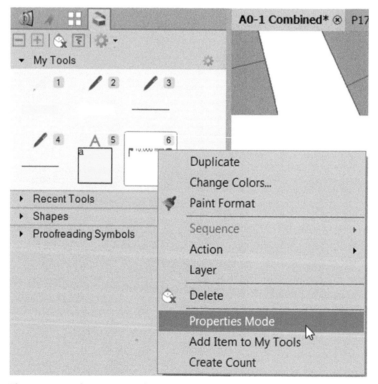

Figure 7-5: Tool Properties Mode

Try drawing another length with the saved tool now. Notice the properties are the same, and the length is new based on the new measurement.

Meet David Gogol

Assistant Civil Engineer, Burns & McDonnell

Photo Courtesy of David Gogol

"When I started at Burns & McDonnell, we were all using Adobe. About a year after I was hired on, the company switched over to Bluebeam, so we were forced into using it. We could still open Adobe if we wanted to, but I like Bluebeam a lot better. It's like an engineering Adobe. A lot of the CAD courses I took in college translate really well into Bluebeam. The measuring tools and the functionality in general in Bluebeam are very similar to AutoCAD. I've been using Bluebeam for two and half years on a fairly daily basis. It's easy to put in markups like text callouts and to ask questions about the content of the design with the drawing review tools. Right now, I'm in the field working on building underground transmission lines, which involves installing PVC ducts and encasing them in concrete. The best thing I like about Bluebeam for this project is the calibrate and measure tool. We work a lot with plan and profile drawings, where the top half of the sheet is looking down at the project, and the bottom half of the sheet is the profile of the design, looking into the earth where our line is going to be. We can use Bluebeam to measure from the top of the grade to the top of the concrete to get measurements on depth of cover and any separation requirements from the exterior of the duct bank. That's really helpful when we cross over existing utilities, because we need to have certain separation requirements between where our duct bank is and the existing utilities. So we can make sure at a high level from the drawings, do we still have our separation here? Approximately how deep are we here?

"When we're having a conference call with the client, someone like me will have the plan and profile drawings up with everything calibrated, so if a question comes up like, 'How far away are we here, how deep are we here, what's our deepest point?' I can quickly get an accurate high-level measurement. So rather than having a drafter there and getting down to hundredths of an inch, we can say, 'We're about 7.5 meters at our deepest spot.' They want high-level numbers, so having that easily and readily available is awesome. It's not down to the hundredths of an inch precision, but most of the time we don't need that precision for questions we're answering in the field. It's really fast and efficient for the work we do to get the general depth here and there. The linear measurement is the biggest tool we use. Sometimes we use the radius tool to get a rough approximation of the radius of bends. All our drawings are scaled so you can go to the scale bar and calibrate the drawing easily. My project uses metric for everything and Bluebeam makes it easy to change from meters to millimeters to centimeters so you can get the necessary precision. I can also measure in imperial units, if needed. For how we use measuring, it's perfect for us. If we were doing this on paper, I'd have to get my scale bar out with the imperial scales and metric scales and round to the nearest whole unit, whereas Bluebeam can get a measurement down to tenths of a meter quickly and easily."

Viewports

Often when a drawing is printed from a CAD program to PDF, a Viewport or two will tag along. Viewports can be great if they were set up nicely by the CAD technician, with the correct scale. This can help save the estimator the step of calibration. However, most of the time, the Viewport is added in automatically and the technician won't be taking the time to set it up with a scale. These Viewports get carried over to the PDF during the plotting process, and will make it impossible to calibrate the drawing correctly. If a user is having trouble calibrating the drawing—a key indicator is that the page turns blue when trying to place a measurement—check the Viewports section of the Measurements tab. If there are Untitled Viewports listed there, select and delete them (Figure 7-6). Viewports can actually be great for setting up two different scales within the same page, but that can be added in later if necessary.

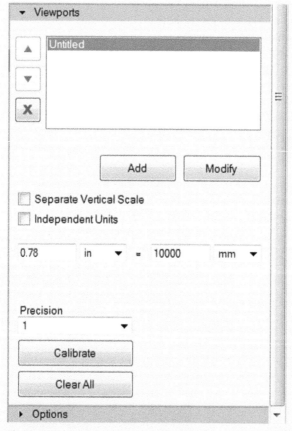

Figure 7-6: Viewports

More Than Just Length

Try out some of the other Measurement tools. Area works in the same way the Polygon tool works, but will calculate the polygon's area based on the calibration. For Area especially, it's helpful to modify the properties with a color fill, maybe adding a Hatch pattern (Figure 7-7).

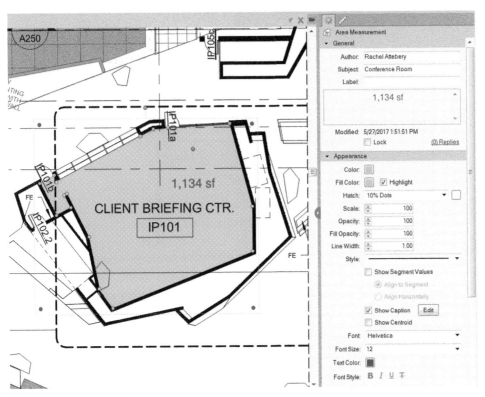

Figure 7-7: Hatch Pattern

For complex areas, it may be necessary to cut out polygons or ellipses from a larger area to get the true area measurement. To do this, first measure the entire area, ignoring the cut-out sections. Then, use the Polygon Cutout or Ellipse Cutout tool to draw a new segment around the cutout sections that will be removed from the larger area (Figures 7-8 and 7-9).

For an area, checking the Show Segment Values box under the Appearance section of the Properties tab will show the length of each side of the polygon. Polylength and Perimeter are similar, except that Polylength shows the length of each straight segment, and Perimeter shows only the total length of the connected segments. Double-click to end the Polylength or Perimeter measurement. The Diameter tool asks the user to start with a point on the circumference of a circle, then pull outward to measure the diameter of the full circle. Interestingly, although there is not an explicit Circle Area measurement, the circle's area and several other measurements can be turned on by clicking the Edit button under the Appearance section of the Properties tab (Figure 7-10).

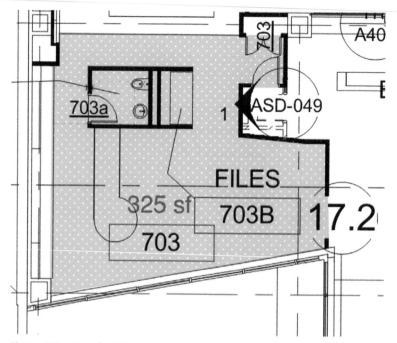

Figure 7-8: Area before Polygon Cutout

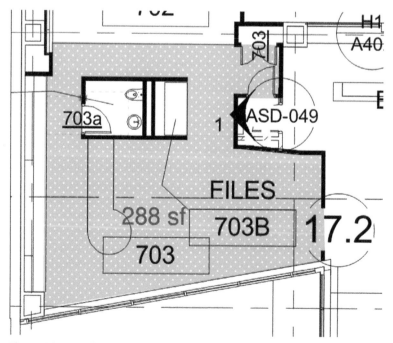

Figure 7-9: Area after Polygon Cutout

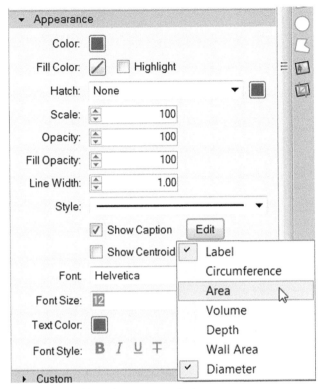

Figure 7-10: Additional Measurement Options

Click any measurements that should be shown on the drawing. These options exist with all other applicable length measurements as well, not just the Diameter tool. The Angle tool measures the degrees of any angle after the user chooses three points. The Center Radius tool draws a circle by selecting two points. The first point is the center of the circle, and the second point is any point on the circumference of the circle. The 3-Point Radius tool also draws a circle, but by selecting three points. Each point represents a point on the circumference of the circle. After the first two points are set, the user can see the radius measurement appear and change as the third point moves around. The Volume tool requires one more step to set up before volume can be calculated. For a given object with length and width and depth, the depth must be set manually before using the mouse to measure off length and width. Click on the Volume tool, then open the Measurements section of the Measurements tab (Figure 7-11). Find the Depth field and enter in the depth of the object. The depth is the dimension of the object that is not visible on a 2D drawing; for 3D drawings, the user can use the Length tool to measure any one of the length, width, or depth of the object, and enter that number as the "depth." The other two dimensions then are captured by the mouse when using the Volume tool. In the image below, the blue 2'-2" measurement was measured by the Length tool and entered manually into the Depth field, and the red 123 cu ft measurement was calculated by clicking the Volume tool and clicking around the face of the 3D object to capture length and width.

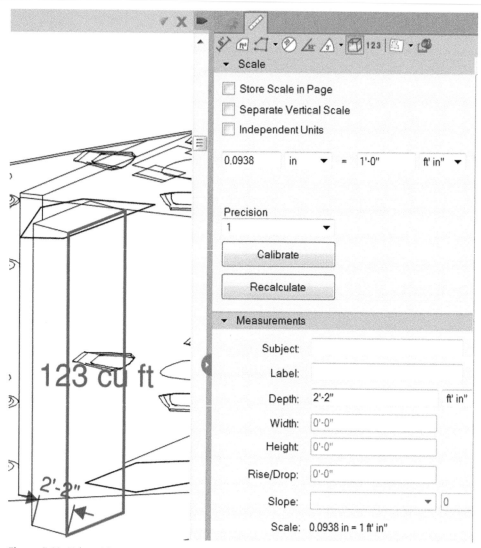

Figure 7-11: Volume Measurement

Count

The Count tool is different than the other tools in that it does not have a dimensional unit assigned to it, but merely keeps track of how many times the user places a certain symbol on the document. The Count tool is useful for counting doors or toilets or trees or any other object that appears multiple times as a distinct entity on a drawing. Set up the Count tool by first clicking on the Count button, then hopping over to the Properties tab (Figure 7-12).

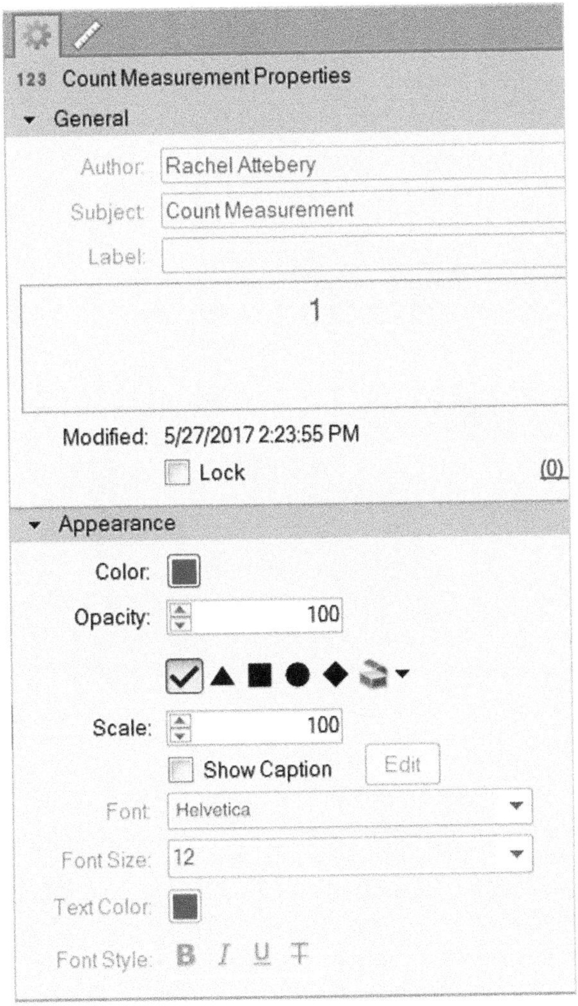

Figure 7-12: Count Properties

Notice the color, symbol, caption, and more can all be adjusted. New with Revu 2017, custom symbols can be used for Count as well. To set up a custom symbol, draw it or take a Snapshot and add it to the Tool Chest. Right-click on the symbol in the Tool Chest and click Create Count. The symbol will now show "1 2 3" next to it, indicating it is available as a Count symbol. Go back to the Count tool and the Properties tab and click on the dropdown menu next to the Tool Chest Custom Count button. The new symbol will appear as an option. After selecting a symbol, the user can start clicking on the drawing to start the count. When in a count sequence, all symbols will be grouped as one markup. To finish counting, hit the Esc key. The user can always pick up again with the same sequence later by right-clicking on any of the symbols and selecting Resume Count. Open the Markups List and notice the count of the objects listed in the Comments column (Figure 7-13).

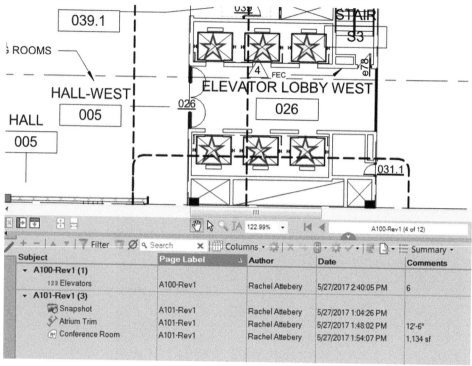

Figure 7-13: Count in Markups List

The user can also resume counting in a sequence by right-clicking on the markups list entry and selecting Resume Count. Sometimes it is necessary to split off some of the counted objects from the sequence, to be deleted or changed. To do this, right-click on any symbol in the sequence and select Split Counts to split off that one object from the sequence. Use the Shift key to select multiple symbols to split off into their own combined sequence. Select the Split All option in the right-click menu to separate all objects into individual counts. It might be easiest to Split All symbols, then use the Shift key to select multiple objects, right-click, and select Merge Counts to combine these symbols into their own count sequence. To delete a symbol from the count, right-click on it and select Delete from Group. Selecting Delete will delete the entire sequence.

EXPERT TIP

Visual Search

Besides manually clicking on each object in a count sequence, the user can use the Visual Search feature to find all like objects and apply a count to them. Open the Search tab in the right-hand flyout, click the Visual radio button, and use the Get Rectangle button to select a snapshot from the drawing. Set whether the Search should look in just this page, the whole document, or even other documents, then click Search. Use the Check All button to select all results, then use the orange checkmark button dropdown menu to select Apply Count Measurement to Checked and choose a count symbol. All found search results will have the count symbol applied and be combined into one count sequence. Visual Search may miss a few objects, so remember to keep an eye out and use Resume Count to grab objects that weren't found the first time around.

Dynamic Fill

Now for one of the coolest new features—Dynamic Fill. This is available in all 2017 versions of Revu (i.e., not exclusive to eXtreme). Click on the Dynamic Fill button in the Measurements tab to pull up the Dynamic Fill toolbar (Figure 7-14).

Figure 7-14: Dynamic Fill Toolbar

Start by clicking on the blue cogwheel Settings button. The first option is Boundary Size. Sometimes the complex region to be measured is not completely closed off by lines. The Dynamic Fill tool only respects lines and will bleed onto the entire page unless it is cut off by a boundary line. The Boundary Size allows the user to choose the size of the boundary line that will be drawn to constrain the fill to a certain region. Fill Size defines how large the starting "paintbrush" is. For a large space, a bigger Fill Size will get the job done faster; for a smaller space, a smaller Fill Size is appropriate. Fill Speed defines how fast the space fills, and Edge Sensitivity defines the line weight respected by the fill. Low Edge Sensitivity will cause the fill to disregard boundaries made by light line weights; High Edge Sensitivity will cause the fill to respect even thin line weights. Click Apply to save these settings. After the Settings are chosen, notice the options next to Create in the Dynamic Fill toolbar. This gives the user the option to choose which measurements are taken by the fill tool. Click on a measurement to turn it on, click on it again to turn it off. Each measurement has a dropdown menu that allows the user to choose an appearance for the measurement.

If the user has already created a custom appearance for a measurement and added it to his or her Tool Chest, that appearance will be available in the Dynamic Fill menu. Recently used appearances are also available, even if the user hasn't specifically added that tool to the Tool Chest. This example will use the same custom appearance for area created earlier, a light blue transparent hatch. Click Apply to set these preferences. Now define any boundaries as necessary. Remember, the fill tool will bleed all over the page unless a boundary stops it. Use the Add Boundary button to draw boundaries where needed. Double-click to end the boundary definition. Now click the Fill button and click and hold in the middle of the region to be measured. The fill will begin to seep out to the edges of the boundaries. In one use of the Fill tool, click and hold multiple times to include all parts of the region that are meant to be measured as one measurement, even if the fill respects boundaries that split the region. When the region is completely filled, click Apply, and watch the appearance change to whatever the user selected (Figure 7-15).

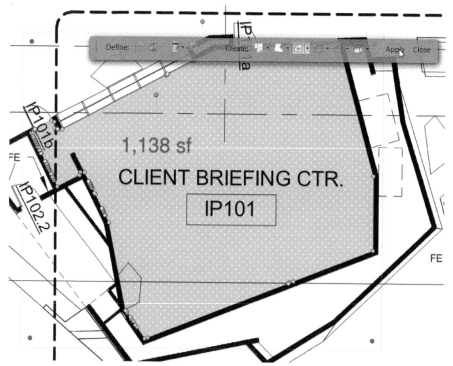

Figure 7-15: Dynamic Fill Appearance

This is a faster method than outlining an entire awkward shape with the Area tool, and is frankly a lot more fun. Even for normal shapes, Dynamic Fill makes a quick job of measuring and eliminates user error of not clicking exactly right around the boundary of the region. Click Close to exit the Dynamic Fill tool.

Sketch to Scale

Sometimes, the user may want to add sketches to the drawing but do it to scale. Once the drawing is calibrated, this is quick to do with the Sketch to Scale tools. The Sketch button lives under the Markup tab, and can also be turned on through the View tab > Toolbars button > Sketch option. Sketch to Scale options include Polygon, Rectangle, Ellipse, and Polyline. Sketch to Scale can be frustrating to use, because the user can't modify the dimensions after drawing the shape; he or she has to enter the dimensions while the shape is being drawn the first time. To do this, take the rectangle as an example (Figure 7-16). Click once to place a corner of the rectangle. The user can hold and drag the mouse and watch the height and width of the shape change with the mouse, but it's hard to get a perfect measurement this way. Instead, after the first click, type in the Width of the rectangle. Then hit the TAB key to move to Height and enter the desired Height. Hit TAB again to specify the rotation of the shape, but this can always be done after the shape is placed. Hit ENTER to finalize the dimensions of the shape.

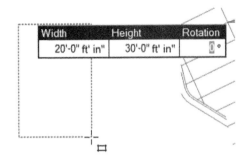

Width	Height	Rotation
20'-0" ft' in"	30'-0" ft' in"	0 °

Figure 7-16: Sketch to Scale Rectangle

For the Polyline and Polygon tools, click once to place the end of the line, then move the mouse in the direction the line should take and type in the length of that segment of line, then hit ENTER. That segment will solidify, and the user can move the mouse again in the direction the next segment should take and type the next segment length and hit ENTER again. When the polyline is complete, double-click to exit the sketch tool and finalize the polyline.

Measurement Options

That's the rundown of the measurement tools available within the Measurements tab, plus some related tools. There are even more options for measuring in the Scale section of the Measurements tab. Store Scale in Page was covered earlier in this chapter, but Separate Vertical Scale, Independent Units, and Precision offer a few more ways to get the most accurate measurements possible.

Separate Vertical Scale: Checking this box will allow the user to calibrate once for the X axis, or horizontal measurements, and again for the Y axis, or vertical measurements. The two calibrations are done separately and can be separate scales. This gives the option to easily capture accurate measurements on drawings with different X and Y scales.

Independent Units: Checking this box allows the user to specify different units of measurement for area and volume than what was used to calibrate length. This is helpful to meet client specifications or unique project units.

Precision: This dropdown menu allows the user to add decimal places to the measurements, for all units except ft'in" and in". For ft'in" and in", precision options are given in fractions of an inch (e.g., ½, ¼, ⅛, ¹⁄₁₆, ¹⁄₃₂).

Rise/Drop and Slope

A few more options exist in the Measurements section of the Measurements tab. New with Revu 2017, the user can specify Rise/Drop for Polylength, which accounts for vertical length in addition to X and Y axis length. For example, if a user is measuring a pipe run on a 2D drawing, from a bird's-eye perspective, it is easy to measure the length of the pipe running backward and forward and left and right. It is impossible to use the lines shown on the drawings to capture up and down, or vertical, length because of the perspective of the drawing. Therefore, if the user knows the pipe travels down 5 feet before turning and continuing to the right, he or she can specify the Rise/Drop of the measurement as 5′-0″, which will add five feet of length to the overall measurement, capturing the true length of the pipe.

Slope is the second addition in Revu 2017 to the Measurements section. Slope allows the user to specify a slope in Pitch, Degrees, or Grade for a Length or Area measurement. This helps account for additional length and area gained by sloping a material. For example, if a piece of pipe runs from point A to point B flat on the ground, it is 5′-0″ long. If a piece of pipe runs from point A on the ground to point B at a 20° slope, the pipe is 5′-3″ long. Slope helps account for that additional length without a lot of extra manual calculations and geometry. Bluebeam helpfully displays the specified Slope along with the measurement on the drawing.

Estimating

Now that calibrating, scaling, and measuring are covered, it's time to get into the Markups List's role to play in estimating. The best way to get the most out of Bluebeam's estimating tools is to set up the drawing right ahead of time. This includes:

1. Creating, labeling, and adding measurement tools to a Tool Chest

2. Setting up Custom Columns in the Markups List

Create Tool Chest

First, create, name, and add tools as described previously in this chapter. Remember to give the tool a Subject label before adding it to the Tool Chest. Remember the right-click, Properties Mode as opposed to Drawing Mode once the tool is added to the Tool Chest. It is helpful to create a whole new Tool Chest and name it something specific, like "World Headquarters Estimating," to keep the new tools straight. This also makes it a simple act to share the Tool Chest with others on the project team. Within the Tool Chest, click the blue cog Settings button and click Detail to turn on the Subject labels for each tool (Figure 7-17). This makes it easy to grab the right tool and start marking off areas and objects. If the tools

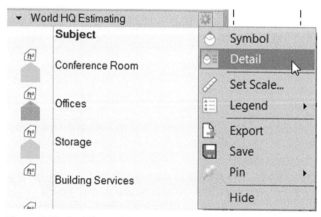

Figure 7-17: Tool Chest Subject Labels

appear too small in the Tool Chest, use the blue + symbol at the top of the Tool Chest tab to increase the size of the tool symbols. This does not increase the actual size of the tools when they are used, just their appearance in the Tool Chest.

Tool Chests can be associated with a Scale. This means if they are used on one drawing or page with a certain scale, they can be reused on a different drawing with a different scale, and they will automatically resize with the new scale. To set the scale for a Tool Chest, click on the blue cogwheel next to the Tool Chest title and select Set Scale (Figure 7-18).

Set the baseline scale. The user even has the option to use the point-to-point calibration tool here, or just manually type in the scale. Once the scale is set, a ruler icon will appear next to the name of the

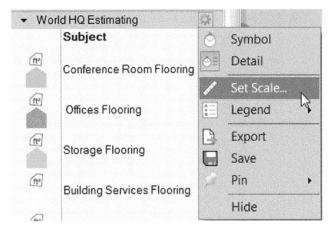

Figure 7-18: Set Scale of Tool Chest

Tool Chest indicating the dynamic scaling is enabled. Click the icon again to turn off dynamic scaling for the Tool Chest. To create a separately scaled area within a page, go to the Viewports section of the Measurements tab. Click Add and draw the area where the new scale should be applied. Define the scale just the same as doing a regular calibration. Test out the separate calibration by measuring a length outside the viewport and then inside the viewport. The two lengths should show different measurements, even though they might be the exact same size markup (Figure 7-19). Try out the scaled tools by placing a scaled markup outside the viewport, then inside the viewport. For a symbol-type markup, the symbol should grow or shrink right before the user's eyes to the scale defined in the viewport.

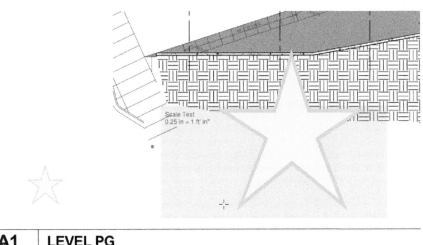

Figure 7-19: Scaled Tool Placed in Two Different Viewports

Create Custom Columns

Second, set up the Custom Columns. This is where the real magic comes in. Open the Markups List tab and click the blue cogwheel for Manage Columns. Click the Add button to add a new custom column. Type a name for the new column. This is the category for materials, where the end goal is to be able to select a certain type of material with an associated cost per unit from a dropdown menu in the Markups List and calculate cost for each measured material in the drawing. "Material" is a fine choice for the name of this column. Click the Type drop-down menu under the Name field and select Choice. Now add some choices of materials. Using flooring as an example, say there are five different types of flooring materials to be used on one floor of a building. Type "Flooring" into the Subject field and pick one of the flooring types to put in the Item field (Figure 7-20). Check the box next to Assign Numeric Value and enter the monetary cost per unit of the material type. Use the same units as the flooring area measurement will use; for example, if square feet is the unit of measurement that was calibrated in the drawing for area, use the price per square foot here.

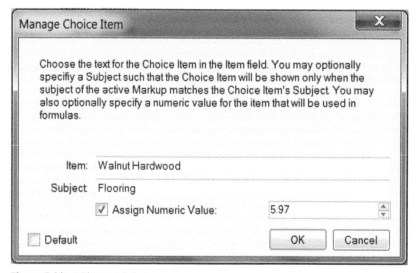

Figure 7-20: Add Material Choice

Add more flooring types as needed. Just like the flooring example, the user can add multiple material types for any category. Beams, piping, appliances, roofing, walls, doors, windows, and so forth would all be considered Subjects, and the different types and costs of each material within a subject would be added as Items with a Numeric Value. Once all Items are added to the Choice options, click the Format button and select Currency from the Format dropdown menu. Specify the number of decimal places shown on the cost and what type of currency to use (Figure 7-21).

Click OK to return to the Modify Column dialog window and confirm all Items and Numbers are entered. Click OK again (Figure 7-22).

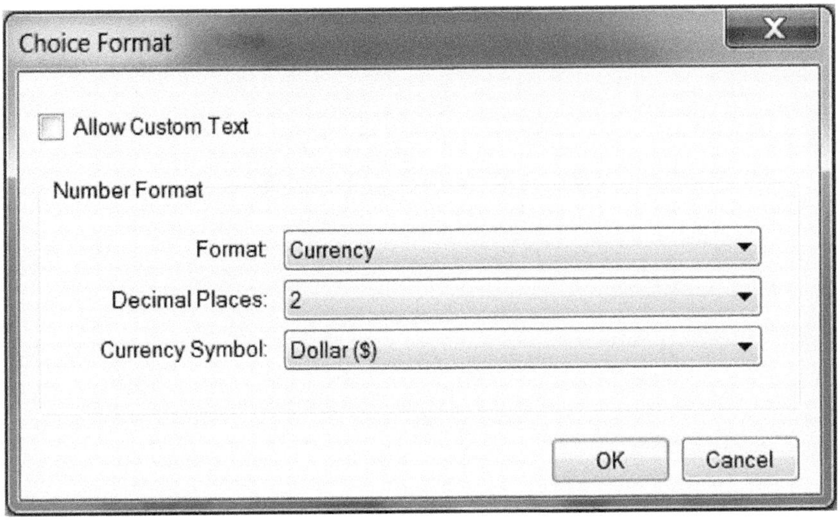

Figure 7-21: Format Material Units

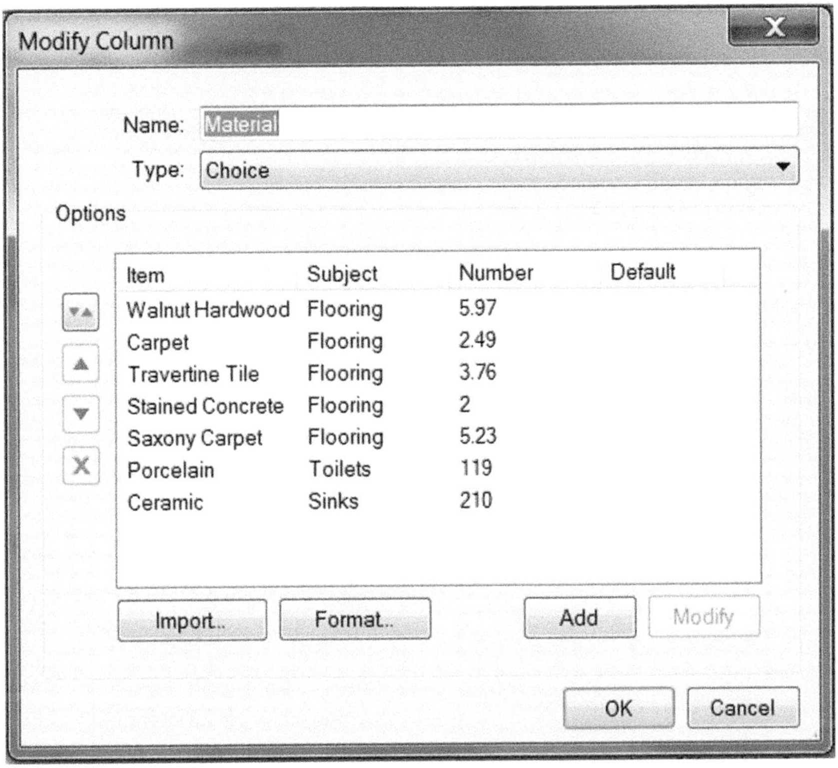

Figure 7-22: Fully Setup Material Choices

Click Save to Profile to save this custom column to the user's Profile. This is useful if the user wants to export and share this Profile. Click OK and watch the new column "Material" appear in the Markups List. Test out the new column by drawing a flooring area using a tool from the estimating Tool Chest. Double-click in the Material column of the new markup's log line item and select a type of flooring (Figure 7-23). The flooring type will be saved in the Material column along with the price per unit of the flooring type.

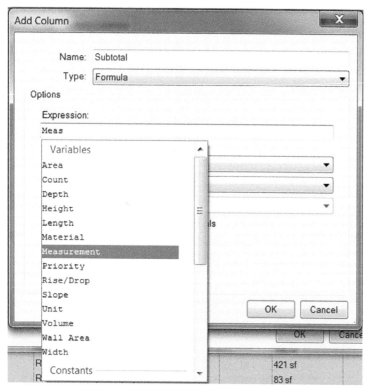

Figure 7-23: Double-click to Select Markup Material Type

Now it's time to add another new custom column for calculating each measurement's cost subtotal. Before doing this, make sure the Measurement column is turned on in the Markups List. Then go back to the Manage Columns, Custom Columns tab and click Add. Choose a name for the cost subtotal column, like "Subtotal." Select Formula as the column type from the Type dropdown menu. The Expression field is where the user will specify that the Measurement column should be multiplied by the Material column to get a subtotal cost for each measurement line item. Use the column names as the variables in the expression field. For example, type "measurement * material" in the Expression field. The "*" symbol means multiply. When the user begins to type in the name of a column as a variable, a list of available variables will appear as options. Double-click on a variable option to use it (Figure 7-24).

In the Format dropdown menu, select Currency, and specify the number of decimal places and currency type. Keep the "Include In Totals" box checked to add the subtotals in this column to a grand total. The user can add multiple "Subtotals" columns if desired, some of which will be totaled and some of which are left out of the grand total. The final column setup should look like Figure 7-25.

Figure 7-24: Variable Options

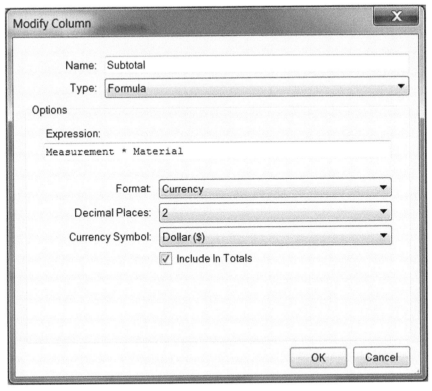

Figure 7-25: Subtotal Column Setup

Custom Columns

This is just one example of how formula columns can be used in Bluebeam Revu. Get creative, and experiment with ways to eliminate external spreadsheets. The Markups List is quite powerful, and since it can be exported to Excel, it is a simple and efficient way to keep information in one place and only export as needed. Any custom columns with numeric values can be used as variables in a formula. The Bluebeam Help menu gives a list of all mathematical operators and functions available in the formula column.

Constants: *e* and *pi*

Functions: *cos, acos, sin, asin, tan, atan, ceiling, floor, ln, log, round, sqrt*

Operators: + (add), − (subtract), * (multiply), / (divide), ^ (exponent), % (modulus), − (negative)

For estimating especially, a combination of custom columns like installation cost, tax, transportation cost, and so on can make it easy to give a detailed, comprehensive sub-total of the cost of building with a certain type of material, all within Revu.

Check out the Subtotal column now for the flooring example (Figure 7-26).

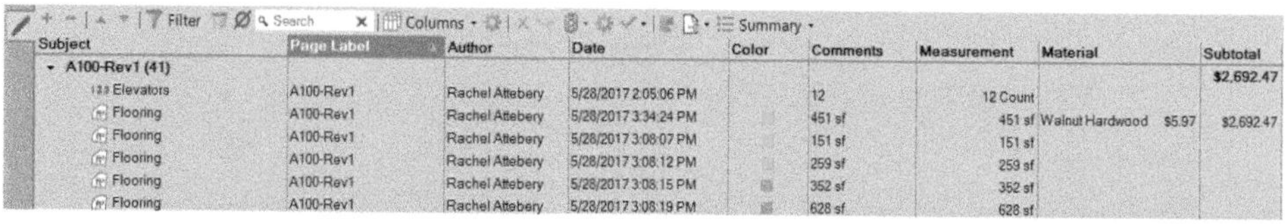

Figure 7-26: Flooring Subtotal Calculated

The Subtotal column has multiplied the 451 square feet of flooring by the $5.97/sq ft Walnut Hardwood cost for a subtotal of $2,692.47. What a great way to find the most expensive room in a building! As other measurements are assigned a material, the Subtotal column will add costs and roll them up to a grand total at the top of each column section. The column sections depend on how the Markups List is sorted. For example, clicking on the Page Label header will sort the log by page label. The Subtotal column will then roll up the subtotals into a total for each page label. The same can be done for Subject to get a total cost for all flooring or all walls, etc.

Now, adding the Material column's items and Numeric Values one by one is kind of like learning how to find the mathematical limit of an equation before learning how to do calculus. It's important to understand the principle of the feature, but a user wouldn't want to add all materials manually forever! There is a faster way. Back in the Manage Columns, Custom Columns tab, click on the Material column and click Modify. The Import button allows the user to import a CSV or TXT file with Item, Subject, and Numeric Value data. Set up the CSV file ahead of time in the following format (no need to add column headings):

Walnut hardwood	Flooring	5.97
Carpet	Flooring	2.49
Travertine tile	Flooring	3.76
Stained concrete	Flooring	2
Saxony carpet	Flooring	5.23
Hollow core	Door	25
Solid core	Door	57

Format for TXT:

Walnut Hardwood,Flooring,5.97

Carpet,Flooring,2.49

Travertine Tile,Flooring,3.76

Stained Concrete,Flooring,2

Saxony Carpet,Flooring,5.23

Hollow Core,Door,25

Solid Core,Door,57

Once the .CSV or .TXT file is set up, click the Import button and navigate to that file. The values will populate into the column choices window (Figure 7-27).

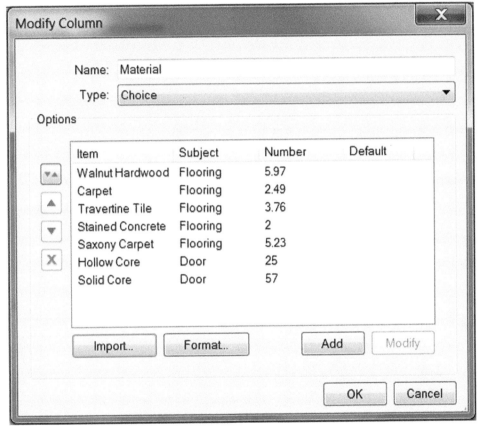

Figure 7-27: Imported Material Choices

This is a fast, accurate way to populate materials, especially when there is a long complex list that is likely to change. Now that the Measurement, Material, and Subtotal columns are set up, the user is ready to start estimating. Using the tools from the estimating Tool Chest, start measuring floor areas and counting doors and whatever else needs to be estimated. Use the Material column dropdown menu to select a material for each measured item and watch the subtotals add up (Figure 7-28).

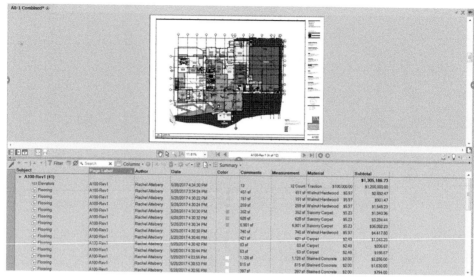

Figure 7-28: Sample Floor Plan Material Estimation

World HQ Estimating		
Description	**Quantity**	**Unit**
☆ Elevators	12	Count
▨ Flooring	1,666	sf
▨ Flooring	2,262	sf
▨ Flooring	6,414	sf
▨ Flooring	7,967	sf
▨ Flooring	8,200	sf
● Sinks	20	Count
▲ Toilets	23	Count
◆ Urinals	8	Count

Figure 7-29: Legend

Legends

Once the materials start to add up, it can be helpful to include a legend on the drawing page in addition to the values in the Markups List. There are two ways to do this. The first way is to click on the blue cogwheel button for the estimating Tool Chest and click Legend, then Create New Legend. This will create a new, dynamic markup with a count of all objects from the Tool Chest (Figure 7-29).

As more counts and measurements are added, the Legend will update. When the Legend is created from the Tool Chest, if any symbols are missing, right-click on a markup with that symbol or properties on the drawing and select Legend, then choose which Legend to add it to. To remove a symbol from the Legend, right-click on any markup with that symbol or properties and select Legend, then choose None (Figure 7-30).

While Symbol/Color, Description, Quantity, and Unit are the default columns included in a Legend, other columns can be shown by clicking on the Legend and opening the Properties tab from the right-hand flyout. Under the Appearance section, the user can click Edit Columns to add more information to the Legend. Other properties like title, font size and color, table style, and more can be adjusted. When adding Gridlines as the Table Style, make sure to increase the Line Width from 0.00 to any positive number, otherwise the gridlines won't appear. The user can use the Source Page(s) field to indicate which pages to pull the legend data from. That way, the Legend can display only data from the current page, or from all pages in a multi-page file. The thing to note here is that each markup can only be counted in a Legend once. So if there is a Legend on the cover page of a set of drawings in a multi-page file, the user cannot add another Legend

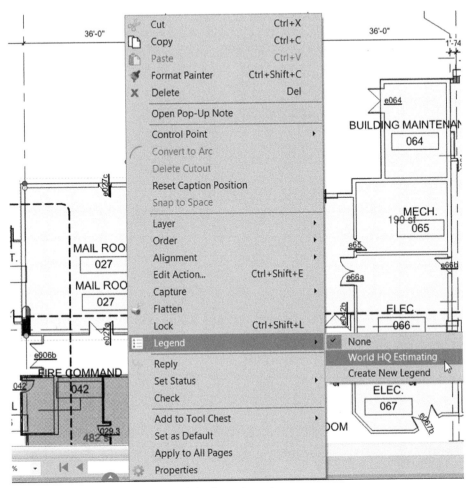

Figure 7-30: Adding or Removing a Symbol to/from a Legend

on each individual page and count the markups a second time on an individual page basis. The Symbol Size can be changed in the Legend to make it easier to see. Margin defines how much white space is given around the Legend data.

The second way to create a Legend is to right-click on individual markups from the drawing and select Legend, Create New Legend. This can take longer but is helpful if the user only wants to include a few types of markups in the Legend, or if all the markups to be included in the Legend are not in the same Tool Chest. Select multiple markups using the Ctrl key and add them all to a Legend at once by right-clicking on any of the selected markups.

Quantity Link

Now for the most exciting new estimating tool of Revu 2017: Quantity Link. This feature is only available with Revu eXtreme. Quantity Link allows the user to set up a custom table in Excel and then create a live link between Excel cells and Bluebeam markups. So, if a table is set up in Excel like Table 7-1, the highlighted cells can be linked to the markups done in Bluebeam and automatically update whenever the user adds or subtracts a markup of that type.

Table 7-1

Flooring	Unit	Cost/Unit ($)	Quantity (sf)	Subtotal ($)
Walnut Hardwood	sf	5.97		=[Cost/Unit]*[Quantity]
Carpet	sf	2.49		=[Cost/Unit]*[Quantity]
Travertine Tile	sf	3.76		=[Cost/Unit]*[Quantity]
Stained Concrete	sf	2.00		=[Cost/Unit]*[Quantity]
Saxony Carpet	sf	5.23		=[Cost/Unit]*[Quantity]
			GRAND TOTAL	=SUM([Subtotals])

Before linking markups to Excel, the user needs to place at least one of every type of markup to be linked on the drawing. It's best to set up the Tool Chest and Custom Columns and perform as many measurements as possible within the drawing first, and then link to Excel. When the user is ready to link, right-click on the first highlighted cell in the table within Excel (Figure 7-31).

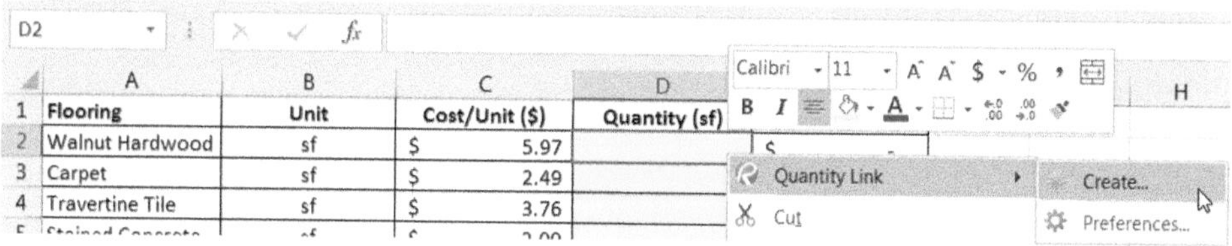

Figure 7-31: Quantity Link Option in Excel

Select Quantity Link with the Bluebeam Revu icon, then click Create. The Select Files Linked to Workbook dialog box will open (Figure 7-32). Click the dropdown menu on the Add button to select multiple files, an entire folder, or an entire folder and subfolders. Just click the Add button to navigate to a file (however, you can still use Shift and Ctrl to select multiple files). Once all the necessary files are added, click OK.

The Create Link dialog box will appear, offering options for which drawing data to link into the cell. Select what kind of Total to use—Length, Area, Volume, Wall Area, Width, Height, or Count. Select what should be totaled—a certain Subject, Color, Author, Material, and so on. For this example, Area will be the Total type, and Material: Walnut Hardwood will be what's totaled (Figure 7-33). Use the second dropdown menu to add filters to what is being totaled. For example, instead of going straight to Area of Walnut Hardwood, the user

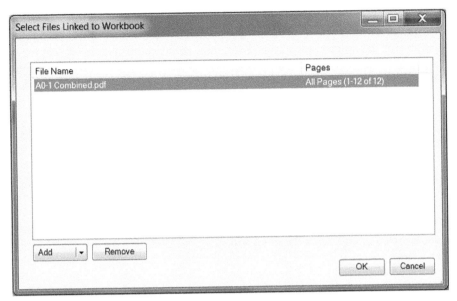

Figure 7-32: Add Linked Files

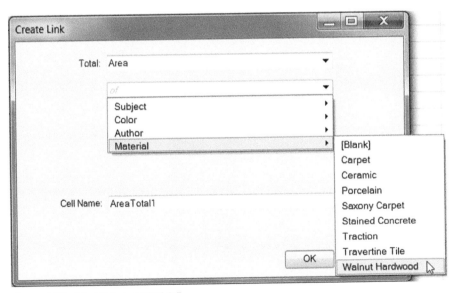

Figure 7-33: Select Data for Linked Cell

could first select Subject: Flooring and then filter down to Material: Walnut Hardwood with the second dropdown menu. If the second dropdown menu is used, a third will appear, and so on until there are no filter options left. Remove totals by clicking the red X next to a total. Give the Excel cell a name if desired; this names the range in Excel but is not necessary to complete the Quantity Link.

Note: The measurement unit will also be imported and populated into the cell to the right of the linked quantity cell, unless this preference is turned off. It can be annoying that the unit imports if the user already has the Excel table set up perfectly according to his or her preferences. To turn off the unit import, right-click on any cell, go through the Quantity Link menu to Preferences. Uncheck the box next to Include Units. Also in this Preferences window is the option to populate the number of markups of this type into the table. This will populate the second cell to the right of the quantity cell, so again, the user can make space for this piece of data to populate, or turn off this preference so it doesn't mess up the user's table. One last preference here—the user can choose to apply a cell fill color to cells that have pulled in live data from Bluebeam. When changing preferences after a cell has already been linked, delete the link and recreate it to apply the new preferences.

The quantity of that type of markup will populate into the Excel table. Test out the dynamic nature of this link by making another mark of the same type on the drawing in Bluebeam. The Excel quantity adjusts immediately after a change is made to the linked drawing. Check out the Markups List subtotals to check against the Excel table. The material subtotals in Bluebeam should exactly match the subtotals calculated in the table (Figure 7-34). If they don't, the user should double-check that the prices for each material type are entered exactly the same in Bluebeam as they are in the Excel table.

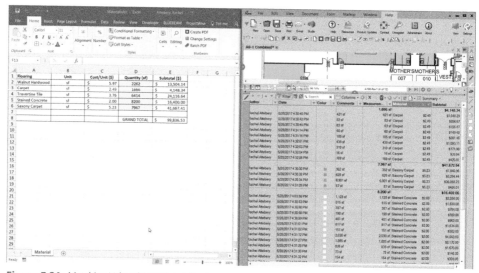

Figure 7-34: Matching Values in Excel and Bluebeam

After the Quantity Links are created, the user can go back and edit them at any time by right-clicking on the linked Excel cell and selecting Edit (Figure 7-35).

The user can add more source files by selecting Source Files from this same menu. Both local and network files can be added for live linking, but there is not an explicit option to select a file from Bluebeam's other integrated document management systems, ProjectWise and SharePoint. This could still be done by adding either of those databases as links within

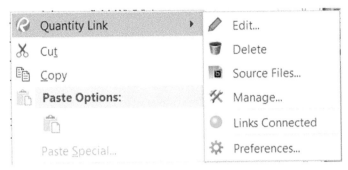

Figure 7-35: Edit Quantity Links

Windows Explorer. The option to link to multiple files allows the user to get a total across multiple drawings rather than just pulling data from one file. This also means that multiple professionals can be placing measured markups on drawings while one estimator manages the Excel spreadsheet that totals measurements from all the drawings. The Manage option enables the user to quickly edit or delete all Quantity Links that exist in the Excel spreadsheet. The Links Connected/Suspended option allows the user to pause the updating of the totals in the Excel table, and then resume it again.

Conclusion

Bluebeam Revu 2017 is an extremely powerful tool for measuring and estimating, whether the user is working with Standard, CAD, or eXtreme. It enables professionals to keep the information in one place with quick, easy ways to distribute it when necessary. While it can take some setup to get the most out of these estimating tools, the results are fantastic. This is clearly a focus area for Bluebeam, so expect more slick features in the near future.

Chapter 8
In the Field

Perhaps the most exciting part of the AEC world is the part that happens outside of the office; where projects get built, boots get muddy, and rubber meets the road. It's the place where problems get worked out, design flaws float to the surface, and the virtual world comes to life.

In today's world, connecting the field to the office is a desired goal and Bluebeam has a number of features and one secret weapon, the iPad Revu app, that can make significant strides to accomplish that.

Even more than the office applications covered by this book so far, field applications are best shared through case studies, stories where Bluebeam made an impact in construction, inspection, or other field applications. Recognizing that, the authors have included three case studies to highlight the significance and broad spectrum of the tool in today's built world. Topics of the chapter include:

- Revu App for iPad
    - Revu and Vu
    - The Graphical User Interface (GUI)
    - Capabilities
        - General Functionality
        - iPad-Specific Integrations and Features
    - Limitations: iPad versus Desktop
- Mobile Access
- Field-Generated Documents

Revu App for iPad

The chapter begins with the iPad app because the iPad has completely revolutionized the construction industry and Revu for iPad has enabled many of the great things Bluebeam users love in a tablet format that is easy to take on the go. Of course there are also some limitations, situations where the iOS platform simply cannot do what the PC platform can do or the tablet horsepower just isn't sufficient enough to run a process that requires heavy computational resources.

Nonetheless, there are features for Revu iPad that are simply great for the purposes they serve.

The Apps: Vu and Revu

As shown in Figure 8-1, a quick search of the App Store reveals that Bluebeam Software, Inc. has developed three apps for iPad: Bluebeam Vu, Bluebeam Revu, and BREAKGROUND. BREAKGROUND is an editorial-type app that was recently introduced by Bluebeam. It focuses on individuals and companies that are pushing boundaries in the industry, and though it is nearly always an interesting read, it will not be discussed further in this text. The remaining two apps, Vu and Revu, are Bluebeam's secret weapons, the way Bluebeam connects the office and the field.

Figure 8-1: Bluebeam Software in the App Store

Similar to the desktop app, there are two versions of Bluebeam for iPad: Vu, the free, lightweight PDF viewer, and Revu, the $10 functional PDF editor.

Though Vu is great for those who only need to see drawings or work inside of a Studio Session, it is the capabilities of Revu that the authors will cover in detail throughout this chapter.

Meet Matt Kossmann, Design Manager, Black & Veatch Special Projects Corp.

As a structural engineer with a passion for field inspection of critical infrastructure, Matt Kossmann's days are anything but ordinary. He's found himself dangling by a rope 200 feet above a gorge in the Sierra Mountains. He's inspected a concrete chamber 300 feet below a 10-billion gallon reservoir. He's been in tight spaces, hot spaces, wet spaces, cold spaces, and dark spaces. He has no fear of heights, he's not claustrophobic, and he has a love for adventure. He also never leaves home without his iPad.

Matt Kossmann Inspecting a Tennessee River Dam

Prior to 2013, Matt would hit the field with an arsenal of equipment: pens, pencils, paper, clip boards, project drawings, a camera, and a variety of measuring tools. Of course that's in addition to his climbing harness, gloves, hard hat, safety vest, and safety glasses. The simple task of organizing and accessing the necessary equipment was complicated by the environments in which he worked.

In 2013, Matt began working on a dam rehabilitation project where he was the principle field inspector. He was tasked with inspecting each of 35 tainter gates on two dams for damage such as bent structural members, cracked welds, excessive rust, and corrosion. The dams were mid-sized tainter-gate dams with concrete spillways and vehicular bridges directly above the gates. For each gate, Matt was suspended from the vehicular bridge and required to maneuver himself around the gate with a series of ropes.

At that time, Matt's typical process was to take hand-written notes on a clip board, record measurements as required, and photograph areas of interest all while dangling. At the end of each day, he would return to his hotel room, review, clarify, and/or correct any of his field notes, pair the photographs he took with the various issues discovered in the inspection, and scan or photograph inspection notes and reports to send back to the home office. Upon returning to the office at the completion of his trip, Matt would create a field inspection report, formalizing his findings in a written submittal to the client.

As such, Matt's days were long and drawn out. They didn't stop when he left the job site. They were also risky from both a data integrity and a professional safety standpoint. If Matt lost just one of the paper copies into the reservoir, he could lose up to a day's worth of work with no way of recovering it. If the office discovered something missing or needed something clarified from the reports, it would be at least the next day before that message would make it to Matt and another day before he could reply or capture the missing piece. At the same time, with so many pieces of equipment tethered to himself, Matt had a lot to concentrate on.

Luckily for Matt, this particular dam project was different than most. The project manager wanted the fastest turn around possible because he saw the potential to complete the entire rehabilitation process in one working season rather than two, creating incredible savings for the client. In the end, the project wasn't completed in one summer, but it did make way for Matt to introduce new technologies into his inspection process.

Following the project manager's request, Matt and his support team went to work exploring the options available. Three solutions quickly rose to the surface, iPads, cloud storage, and Bluebeam. Recognizing cloud storage could be done using Bluebeam Studio, the final path was as simple as using the low-cost Revu app on an iPad.

Matt Kossmann Inspecting a Missouri River Dam

For Matt, the combination met several needs. First, he could store and manage all of his construction drawings, specifications, and field notes in one place. They would always be accessible whether he was hanging on a rope, in the work truck, or at his hotel room. Additionally, the documents could be hyperlinked, searchable, and bookmarked to ease Matt's efforts in looking for information. When dangling from a rope, every convenience counts.

Second, it provided data security and stability. So long as Matt was connected to the Internet, he could push his inspection data to the cloud periodically or even continuously, keeping it safely out of catastrophe's way. Though Matt could still lose the iPad, doing so would not simultaneously lose the data. Considering professional rates, one day of lost data greatly exceeds a lost iPad.

Third, it provided the rapid response the project manager was looking for. With the more frequent uploads, the office staff could have access to the inspection data within minutes or hours instead of days. Additionally, it created a direct connection between Matt and his support staff back in the office. The two could now communicate via text message, e-mail, or even Face Time all day long. When a question came up, Matt could actually *show* the office what was happening with a live, streaming video.

Fourth, keeping electronic records significantly reduced the follow-up time for Matt. He no longer had to return to the hotel and go through the time-consuming process of cleaning up and digitizing the report. Matt also discovered that if he took care in creating his notes, he could essentially use Revu's summary tools to create a very nice report for the client with less than an hour's worth of additional work, again significantly reducing the time Matt spent on the back end.

Fifth, the setup could easily tie pictures to issues. Using the iPad as his camera, Matt was able to directly embed and attach photos to the notes he had taken. If he identified a crack in a weld, he simply attached a photo of the crack to that note, keeping the two nicely linked and organized. In the traditional method, matching photo numbers to issues he identified was a painstaking nightly effort.

Sixth, it reduced the number of devices Matt had to manage. The iPad with Revu replaced his camera, construction document set, field notes, clipboard, and pen/pencil.

Of course the change didn't come without hurdles and concerns. Battery life, glare, durability, and signal were among them, but Matt was able to manage all four.

As he began using the iPad, he monitored the battery level carefully, charging over lunch the first few days until he became comfortable with the pace at which the device burned energy. Soon he recognized he could generally function for a full day on one charge.

He found that glare was hardly ever a problem for him. Because he was more often than not hiding in the shadows of the structure, the sun was generally blocked and glare was simply no issue whatsoever. In cases where he was in the sun, he found that it wasn't too burdensome to maneuver the iPad to a position that allowed him to read it.

Durability was solved using a heavy-duty OtterBox Utility case with field accessories such as a lanyard, leg strap, handle, and rain cover. Matt has been in some of the roughest conditions of any inspector and though his case shows signs of these experiences, the iPad inside is in as good a shape as day one.

Rugged OtterBox Case Protector

The one item that did need resolution was the signal. At the onset of the project, the team aimed for streaming data, continuous uploading of all of Matt's field notes, but that required a continuous and reliable internet connection. Matt found that as he moved around the tainter gates, he would temporarily lose signal when he was behind large steel plates, or hidden in a concrete corner. The constant reconnecting slowed him down significantly and he quickly shifted to periodic cloud uploads every hour or two.

Overall, the technology on the project worked fantastically well, and before it was complete, Matt's team had the contractor and subcontractors using Revu also, sharing documents in real-time fashion like never before.

Today, Matt continues to use his iPad for all of his inspections. In fact, he wouldn't have it any other way. The time he saves in creating reports for the client and the quality of those reports is so great that it easily balances any benefits of the old way.

"I can tell you that often my field contacts download Revu on the spot after we talk about the app and I show them the functionality." Matt said. "Recently, one high-ranking client even asked if he could get an official copy of Revu because he thought it sounded way better than what he was used to."

To Matt's benefit, the Revu app and his experience with the software have continued to grow over the years since that first project. He has discovered hacks, new tricks, and capabilities to save time, be more organized, and produce a higher-quality product.

To name a few, the way in which Bluebeam handles photos has been significantly improved. The introduction of Capture made it easy to organize photos with markup comments, and now it's also easy to export those native image files and attach them to a summary PDF. The shutter button got relocated to allow single-handed firing, a small change that made a big difference. Batch Summary gave Matt the ability to report on multiple documents and format the report in a more professional way with company letterhead, headers, and footers. The signature tool made it easy to sign off on anything as needed.

Matt continues to feed suggestions to Bluebeam and they continue to answer the call, making valuable upgrades at each version release. What was his latest request? He wants a flash for the camera. So, Bluebeam developers, if you're reading, can Matt get a flash on his camera?

Matt Kossmann Inspecting an Underground Tunnel System

When field assignments arise or there's a need to be mobile, don't forget the power of Revu for iPad. No doubt it's not as powerful as Bluebeam for PC, but Matt's story clearly demonstrates that it can still have a big impact.

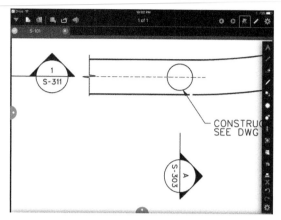

Figure 8-2: Revu iPad Graphical User Interface

The Graphical User Interface (GUI)

The basic Revu iPad interface is shown in Figure 8-2, and as the reader can see, it looks very similar to Revu for desktop. There is tabbed navigation, there is a simplified menu at the top, there are markup tools on the right margin, and there are blue semicircles on the left and bottom sides where Revu's panels and tabs are hiding. (Note that there is no semicircle on the right side because there are fewer tab options available as the iPad most likely has less screen real estate to work with.)

Highlighted in Figure 8-3, there are six buttons within the simplified menu bar at the top left corner of the GUI. The first button on the left is the downward-pointing orange arrow, which opens and closes the left panel, revealing the five available tabs: Tool Chest, Thumbnails, Bookmarks, Search, and Measure. The second button is the New Document button, which does exactly what you expect: creates new PDF documents. Third is the Recent File button, which also performs exactly like Revu for PC. Fourth is the Document Management button, which allows the user to browse for files and save files in various locations, whether on the device or in some type of cloud storage system. The fifth button is used to export files to other people or storage locations. And finally, the last button is the Studio button, which has been given its own chapter within this book.

Figure 8-3: iPad Upper Left Menu

As shown in Figure 8-4, the upper right menu has a total of five buttons. The first two buttons on the left are simply for navigating the document by pages. The third button toggles the markup tools, again likely to allow the ability to save screen real estate. Fourth from the

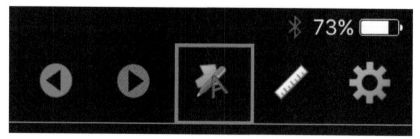

Figure 8-4: iPad Upper Right Menu

left is the toggle for the measure toolbar, which brings the various measuring tools into the markup tool bar on the right. And finally there is the properties button, which is very similar to the desktop app, but has fewer submenus available.

Just like default Bluebeam for desktop, the markups list is at the bottom and may be opened by clicking or sliding the blue semicircle. As shown in Figure 8-5, the markups list appears less detailed, but that is not entirely true. Using the properties gear on the top right side of the lower panel, the user may turn on any data columns that exist within a given document. Unfortunately, he or she cannot add new data columns at this time, but perhaps this will be possible in the future. The user will also notice filter icons allowing all the filter capabilities of the full desktop application.

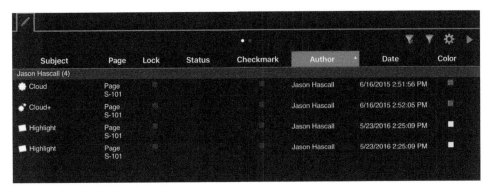

Figure 8-5: Markups List

One small nuance the reader should notice is the existence of two small dots in the top center of the markups list. One is white and the other gray. When the columns in the markups list are too many or too wide to fit on a single page, Revu allows the user to swipe left or right to see the rest of the columns; the more dots on the top, the more pages of columns that exist. Alternatively, the user can utilize the right-pointing, blue arrow adjacent to the properties gear to move to the next page of columns.

EXPERT TIP

Properties for the Markups

Like the desktop app, there is a properties gear button that allows the user to edit the properties of his or her markup, but at the very bottom of the markup menu, it's not nearly as easy to find as it is on the desktop app.

Clicking the icon, the user will notice that what was the markup menu has transformed into a unique properties menu. Unless a markup has specifically been selected, the icons are gray and unavailable. As soon as a markup is available, the icons will come to life.

The user may return to the markup face of the menu by clicking the markup icon that has replaced the properties icon at the bottom of the menu.

There are two other good features to note about the markup menu.

First, it is a scrolling menu on both the markup side and the properties side. The user may simply touch the menu and slide it up or down to find the desired tool or property.

Second, the undo, redo, and delete icons may be found at the bottom just above the properties gear.

Capabilities

Many beloved features of Bluebeam Revu for PC also exist in the Bluebeam iPad app. From tool sets and measuring capabilities to Studio access and searching, the iPad app is a very functional version of Revu. In general the equivalent tools are fairly intuitive to learn and understand.

General

The most significant hurdle for the user is typically adjusting to the device interface. The mouse is, of course, replaced by touch and the keyboard is essentially restricted to typing. As such, the functionality enabled by keyboard shortcuts, right clicks, and combinations of the keyboard and mouse are essentially defunct. The good thing is that in many cases these functions have been replaced by something just as simple.

The touch controls can be sensitive and certainly take some practice and finesse. However, once the skills are mastered, a user can do much of what he or she is capable of doing with the mouse. For instance, holding the touch often results in the equivalent of a right click. As an example, a user may right-click on a markup by selecting it with a single touch, then simply touch and hold. The user will see a menu appear, at which point he or she can release the touch and select from the menu. Figure 8-6 shows a simple example of the right-click options for the cloud mark displayed.

Within this particular menu, the reader can see seven options, which typically show up under the right click menu on Revu for PC. To exit the menu, simply make a selection or click anywhere else on the screen.

The keyboard, of course, works similarly to other apps for iOS, appearing from below when the action requires typed text. To hide the keyboard once again, simply click outside of the text area or click the button in the very bottom right-hand corner.

Revu for iPad also has several features and capabilities that Revu PC does not. Largely, these features lend themselves well to tablet-type devices and do not work well on PC. A few of those are highlighted below:

- Signature Tool: The signature tool is definitely the best way to capture a professional signature. Not only does the tool create a vector markup of a signature, it provides

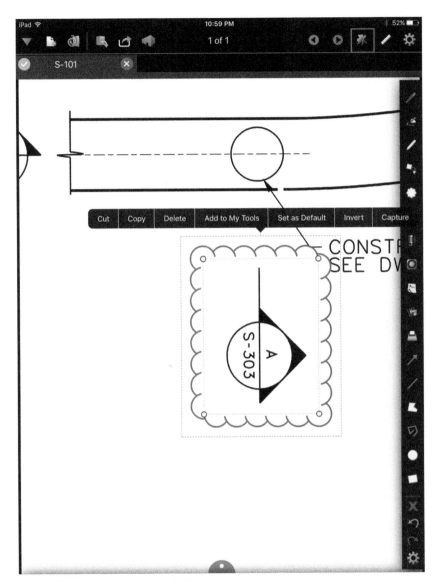

Figure 8-6: Touch Features—Right Click

an interface that is simple for the user to work in. Figure 8-7 shows the tool in action. The signature tool is accessed by clicking the tool icon that resembles a hand signing on an identified line. Once selected, the user simply draws a line at the location and length where the signature is desired. Immediately, a signature pop-up box appears and the user is invited to "Sign Here." Once complete, he or she may select OK and the signature will be scaled and placed in the document as a markup.

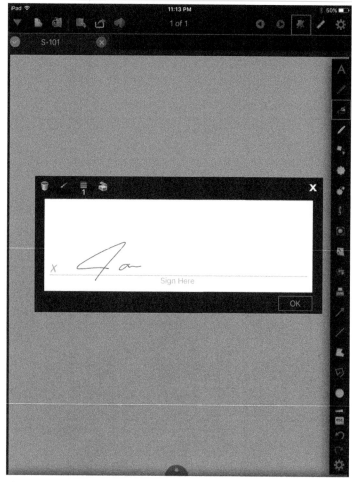

Figure 8-7: Revu iPad Signature Tool

Sign Like a Pro

The signature tool may seem like a silly tool to highlight. After all, are there any significant problems with utilizing the pen tool and creating a signature directly on a PDF? The answer is no. The final product will be essentially the same in both scenarios; vector content with adjustable color, line thickness, and scale.

The great thing about the signature tool, however, is the small window it creates for signing. When signing in the full paper space of the screen, the user must be cognizant of touching the screen with parts of his or her hand because Revu will register a mark at each location where a stylus or a part of the hand comes in contact with the screen. Signing an iPad without resting a hand on the screen is very difficult.

On the contrary, utilizing the signature tool blocks out the whole screen except for the signing window in the center, thus allowing a user to touch the screen anywhere outside of the box without impacting the resulting signature.

■ Voice to Text: Harnessing the iOS voice to text capabilities, Revu for iPad can save a user significant amounts of time in typing on the tablet. Instead of typing a comment, the user only needs to touch the microphone button at the bottom. After the prompt tone, anything the user speaks clearly will be recorded in text in place of typing. Voice to text in action is shown in Figure 8-8, with the voice recognition window at the bottom and the text appearing instantaneously in the selected text box.

Figure 8-8: Voice to Text Feature

■ Voice Memo: Sometimes a message is too long to type or may be too detailed to be written. For these cases, Revu developers added a voice memo option. By clicking the microphone icon, the user may record a message and embed it in the PDF file. When the user shares the PDF file with others, the voice memo can be replayed for the recipient to listen to. Figure 8-9 indicates the location of the voice memo tool.

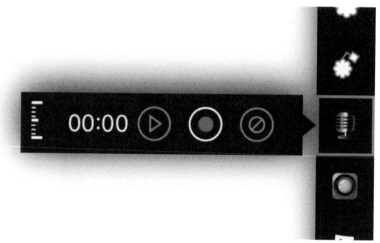

Figure 8-9: Voice Memo Button

■ Camera Integration: The camera integration is surely one of the most widely used tools in Revu iPad. Photography is commonplace in the field environment, so Bluebeam developers made it easy to use. For taking and embedding images directly within the PDF, the user may select the lens icon from the tool menu. A full-screen window will open, as shown in Figure 8-10, and the user can shoot the desired photo by clicking the button on the right. A few features to note are the ability to zoom in/out with the slider bar on the left, the ability to reverse the camera to "selfie-mode" instead of forward mode, and finally the shutter button being located on the right to allow for easy one-handed operation. The camera is also integrated with the Capture feature, which can be accessed by right-clicking (touch and hold) on a markup. The greatest benefits of capture are that the photos do not clutter the page and that photographs may be electronically linked to a comment such that the comment may be further explained with images. Additionally, the images in a Capture are easy to browse and easy to export.

Studio

Chapter 4 introduced readers to Studio, Bluebeam's online cloud collaboration tool where multiple users can markup a single PDF file at the exact same time. The good news is that this incredible functionality is also available on the iPad app. Both Studio Sessions and Studio Projects can be accessed from the iPad, making Studio a great tool for collaborating with the field.

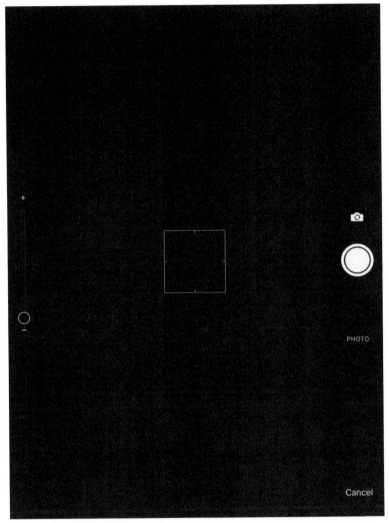

Figure 8-10: Camera Tool

While the desktop app seems to separate Studio from the traditional Windows folder structure, while using the iPad app, users will feel like Studio is just another folder or location for storing files. Because it is possibly the most difficult piece of navigating the iPad, the folder structure and storage locations will be discussed in detail later in the chapter.

One important feature that was introduced in Revu iPad Version 3 was Studio GO. Often large PDF files took significant time to render content, slowing down workers in the field. Studio GO is the Bluebeam developer's solution to this issue. When using Studio GO, files render almost instantaneously as demonstrated in this video. www.bluebeam.com/us/products/revu/2015/ipad3.asp.

Studio GO can be enabled by touching the properties icon in the top right-hand corner of the Revu window, selecting Studio, and then toggling the Enable Studio GO slider, as shown in Figure 8-11.

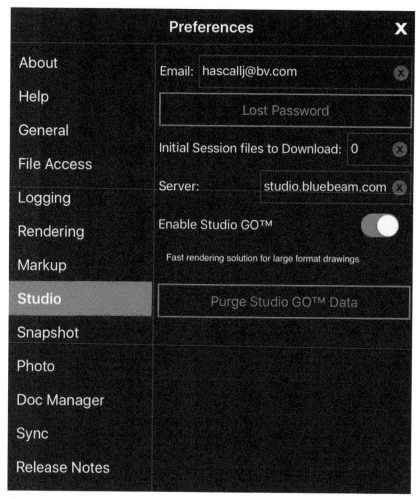

Figure 8-11: Studio GO Toggle

Offline Mode

As colleagues and partners with early users of the iPad app, the authors heard numerous issues with Studio access for iPad. Unlike office internet connections, field connections are often less reliable, sometimes changing by the minute.

One example is when a field inspector has to be moving around a large inspection item like a tower, platform, tunnel, gate, or other industrial structure. These structures often interfere or block the cellular signal, causing the iPad to drop connection.

In the early days of Studio, this caused frustration for users because they were required to continuously log in and out of Studio. This was so burdensome that most chose to avoid Studio at all costs.

Luckily, Bluebeam developers resolved that issue with Offline Mode.

Using Offline Mode, the user is able to work within a Studio Session document just as if he or she was connected; however, the markups aren't posted live but are kept as "Pending" until the iPad connects again. Though this doesn't result in the instantaneous collaboration typical of Studio, it does alleviate much frustration from remote field users.

A similar behavior can be used with Studio Projects, where project files may be synchronized locally to the device. That copy can be viewed and edited and then used to update the server copy when reconnected.

Limitations

While many features Revu users know and love exist in the iPad app, there are still a significant number of features that do not currently exist in the iPad app. Some of those features are likely in development for iOS. Some of those features will likely never be added due to the use case applicability. And other features will likely never be added due to the operating system.

In this section, some of the limitations will be discussed, along with any workarounds known by the authors.

Document Functions

The first set of limitations is in the PDF document editing capabilities. It's very clear that the app creators never intended Bluebeam for iPad to be a PDF workhorse. Instead, they've focused on the workflows and use cases for the mobile user, assuming the heavier document editing functions can be performed by an office counterpart using the full PC version. Most of these features have been explained in chapter 2, while others have been discussed throughout the text.

A few features that have no iPad equivalent functionality are:

- Split Document
- Reduce File Size
- Rotate Pages
- Compare Documents
- Document Security

The iPad app is simply not capable of performing these functions. Similar document functions do have some functionality on iPad, though it may be limited. Consider Extract Pages, for example. The full PC version has a very robust Extract Pages feature, allowing the user to perform all kinds of edits on the pages as part of the extract function. On iPad, the capability exists, but in a very limited fashion. Users can simply e-mail a page or selection of pages from a document.

To do so, from the open document, select the orange down arrow in the upper left and choose the thumbnails tab, as shown in Figure 8-12.

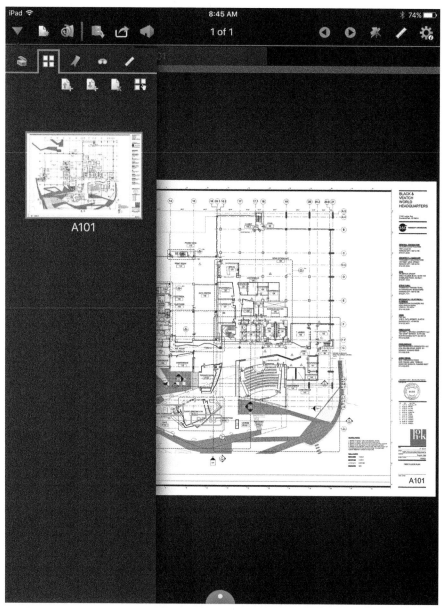

Figure 8-12: Thumbnails Tab

Next, touch the thumbnail select button at the right side of the tab and tap each page that should be included in the selection. In the example, the file consists of only one page and therefore only one page has been selected.

As the pages are selected, an orange border appears around the perimeter, indicating those pages have been selected. Simply touch the page again to deselect it. If the user would like to select all the pages, simply touch and hold any page in the list and a menu option to "Select All" will appear. The same process may be followed to deselect all.

Once the selection has been made, touch and hold one of the selected pages and the user will be presented with a pop-up menu with the "Extract and Email" option, allowing the user to e-mail the selected pages through any e-mail account linked to the iPad. Figure 8-13 shows the export menu option.

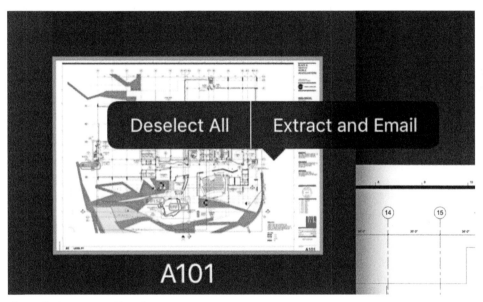

Figure 8-13: Extract and E-mail

Additional document type features that do have some functionality in the iPad app are the ability to add pages, delete pages, create new documents, and reorder pages.

Digital Signatures

Because an earlier section of this chapter noted the capabilities and uniqueness of the signature tool, it's extremely important to clarify that the signature tool is not creating a digital signature as discussed in Chapter 6.

As noted previously, the digital signature is a special type of electronic signature that includes digital encryption and document security features, none of which are being applied by the iPad's signature tool.

Sets

Chapter 5 highlighted, and encouraged, the use of Bluebeam Sets as a great way to manage documents throughout the design and construction process. The unfortunate truth is that the iPad app is not Sets capable. The good news is that developers have provided a very functional workaround and it has to do with the Drawing Log introduced with Tags in the launch of Revu 2016.

Referencing Chapter 5, the reader should recall that the Tags functionality has the option of creating a hyperlinked Drawing Log listing all the files in a given set in a manner that hyperlinks each of their titles to the individual files. For iPad users, this Drawing Log is the key that unlocks much of the functionality of Sets in an iOS environment that doesn't support those features.

Once the Drawing Log has been exported, the iPad user may simply open the log and use it to select any sheet in the Set. He or she will not be able to scroll through the pages of the set the same way PC users do, but keeping the log open as the first tab in the Revu window gives access to any of the sheets with a simple touch. See Figure 8-14 for an example of the Drawing Log being used on Revu for iPad.

With the help of the Drawing Log, Bluebeam Sets become almost as beneficial for iPad as they are for PC. There is no longer a need to open extremely large files, which often slows down or even crashes the Revu app.

The Rest

Not surprisingly, many of the higher-order features and capabilities of Bluebeam for PC are not available on iPad. The processing horsepower simply isn't available on the slim tablet. The good thing is that most of these features aren't conducive to field applications.

Two examples of these features are Batch Functions and OCR. In both cases, these features are intensive processes that utilize a significant amount of memory. Luckily, they're also features that can be executed before the documents hit the iPad app. Once an OCR has been completed, the text stays searchable and that is a capability of the iPad. The same is true for Batch Functions like Batch Link. Once the hyperlinks have been created, they will function on iPad exactly as they do on PC.

Mobile Access

One of the greatest capabilities of the Revu app for tablets is the ability to access project data.

Consider for a moment the time wasted in a single construction professional's day of work. How many minutes does it require each morning to determine which documents he or she will need for the day? How many minutes does it take to print out those documents? How many copies of documents are discarded after a single use? What happens when the worker reaches the 14th level of a jobsite only to realize he or she didn't have what was needed?

Field tablets eliminate all of that, providing access to any document, at any time, in any place, with a simple internet connection. No need to print documents each morning, no need

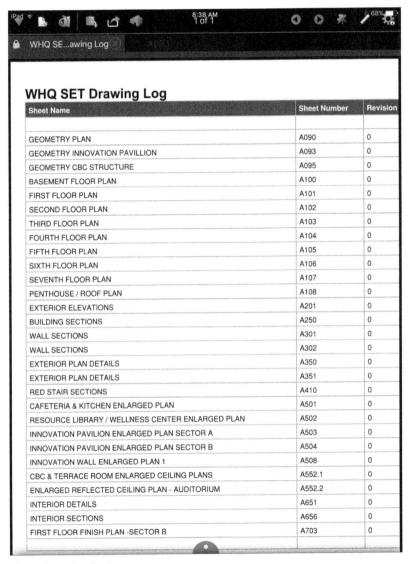

Figure 8-14: Drawing Log

to anticipate what prints will be required on the 14th floor, no need to carry thousands of sheets of paper in a work truck.

It's true that Bluebeam Revu isn't necessarily required for this functionality, but it's also true that the Revu app is making the hard stuff easier. With the Recents tab, Studio, and Batch Link, finding documents is no longer the arduous task of digging through haystacks. Instead, it's often a simple touch of the screen. It can even be taken further into a type of dashboard or toolbox, organizing documents in a very visual way as highlighted in one of the case studies at the end of this chapter.

Field-Generated Documents

While this chapter has clearly noted that it's possible to contribute to a PDF document, the focus has largely been on accessing PDF documents and the capabilities of the iPad. It hasn't yet significantly explored the uses or abilities to generate documents in the field.

CASE STUDY

Meet Vanessa Taylor, IT Special Projects, and Sarah Garcia, Corporate Trainer, P1 Group, Inc.

Vanessa Taylor and Sarah Garcia work together at P1 Group, a national contractor that specializes in mechanical, electrical, and plumbing construction, fabrication, facility maintenance, and energy services. P1 Group has over 900 employees. Although they work for an MEP contractor, neither has installed an HVAC unit nor wired a building. Instead, these women stand out as enablers who work behind the scene to make P1 Group's operation a well-oiled machine.

Through training, scripting, smarts, and creativity, they assisted in creating an automated process that is second to none. It makes P1 faster, it makes P1 better, and it makes P1 stand out in the crowd.

With national reach and hundreds of professionals, managing day-to-day work efforts was a logistical train wreck. Coordinating, organizing, storing, and sharing the documents and reports has always been an important practice for every contractor and P1 was no exception. However, as the organization grew and the work became more distributed, keeping track of everything in a respectable way grew difficult.

P1 Field Professional Utilizing iPad

P1 Professionals Collaborating in the Field

P1 Group, Inc. first implemented a SharePoint project management system to establish consistency. A SharePoint workspace is generated for each project and project documentation is created using templates and workflows. The SharePoint system streamlined processes and created consistency for the office. However, field professionals had trouble figuring out where to store their data, what to call their data, and what exactly their data should be. In addition, paper copies of plans quickly became outdated and the chances of installing the work incorrectly were heightened. The leadership team at P1 could see that these problems were beginning to hinder them. From a legal and good business practice standpoint, something had to be done.

They called on Sarah and Vanessa to develop a method to share project information with the field. Sarah Garcia is a corporate trainer for P1. She works with P1 leaders to develop procedures, then coordinates and contributes to all of the training for the P1 project managers and field professionals. Everyone at P1 Group has learned something from Sarah.

Vanessa Taylor, on the other hand, is an IT mastermind. She helps manage the P1 servers and handles all of the document storage and security needs. And together with other P1 employees, she has been part of creating a solution that meets most of P1 Group's needs.

When the project first began, P1 had around 780 field-based employees. With that many minds working simultaneously, consistency was an obvious struggle and data forms were the obvious solution. The team created PDF templates for daily logs and short-interval plans.

The idea was that with the available templates, each professional would begin to process project documentation in a like format, improving consistency between projects.

The second hurdle the team needed to overcome was how the field team received the templates and where they stored them once complete. Like many companies, P1 had the struggle where different professionals stored information in different places on different jobs. In turn, it was difficult for office personnel to find the information they needed when they needed it.

The development team wanted a way to automatically distribute these required forms to the professionals who needed them and automatically recover the forms when complete. They imagined a data "push" to hundreds of iPads followed by a data "pull" to capture the field inputs.

Was there a way to do this without a web of e-mail messages that had proven to get lost, overlooked, deleted, or otherwise ignored?

The answer was yes. And the team set out on a course to make their vision a reality. Step 1, hire a developer to help.

Over the next seven months, P1 Group laid out the vision and developed the solution, which can be summarized something like this:

- Once each week, blank copies of the daily logs and short-interval plans are semi-populated and pushed to the field professionals' iPads.
  - Project information is pulled from a database and entered automatically.
  - Multiple daily forms are created, one for each day.

Sample Form List for Multiple Users

- When documents are opened on the iPad, the cached copy is updated or the user can choose to sync manually files to the iPad.
- Certain events or entries trigger automatic actions and document routing.
  - Documenting an accident, injury, or near miss on a daily log triggers a script that creates and sends an e-mail to the project managers and the safety department.
  - Damaged freight triggers an alert to the project managers to make them aware.

So how did they do it?

They began with the combination of Bluebeam Studio Enterprise and SharePoint, both utilizing a SQL database.

SharePoint serves as the hub for all projects, it's where everything is stored and where everything is sourced. When a new project is secured, the project team enters the relevant project information such as location, client, and so on into a Job Board database. Once the project information is entered, a job number is assigned and then a workflow runs to set up a SharePoint site specifically for that given project. P1 can then manually kick off a SharePoint workflow, which creates a Bluebeam Studio Project, applies permissions, and grants access to each of the project team members.

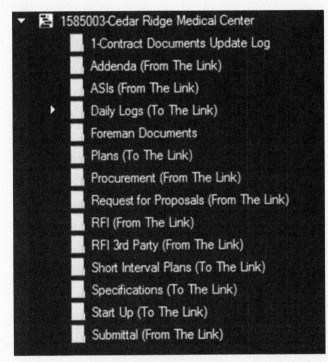

Sample Studio Project for Cedar Ridge Medical Center

The project setup process also includes workflows that create all the necessary field forms for the project, based on the inputs, and populates the respective fields with information from the database, such as client and project name.

From that point on, scheduled workflows keep the project teams up to date. Once a week, the script moves the previous week's completed forms to a completed folder and creates a new week's forms. All the files are named appropriately, all the files are populated consistently, and all the file permissions are set up as needed. With a cellular connection on each of the iPads, files are checked out and updated by the field professionals. When they are checked back in, they are automatically synced to the Studio Project.

The sync also includes any updates to project information like drawings or specifications, so each field professional always has access to the latest and greatest.

Finally, the SharePoint workflow goes one step further and issues alerts via e-mail for certain triggers that require the attention of a manager.

P1, with the help of a developer, had accomplished what the team set out to do. They had developed the software, but they still had to make it stick, and that is where Sarah played the largest role. The company recognized that changes require training, and this change was a big one.

Sarah quickly developed a training program that would be applicable to all field personnel who would be using the iPads and the new system. They would visit the home office and learn the entire system in about 6 hours. Follow-up training is often conducted at the jobsite in the environment where the iPads are used.

To date, the P1 system remains the most sophisticated and beautifully simple autonomous document management setup the authors have ever seen. It serves the purpose for P1 and they're even starting to see some extra benefits with the use of existing Revu features like dictation. Who'd have guessed that site foremen would prefer to talk instead of typing?

P1 BIM Boxes for Large Screen Viewing

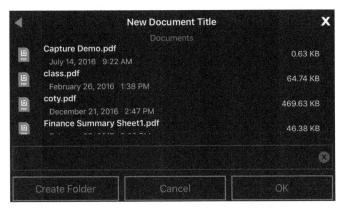

Figure 8-15: New PDF on Revu iPad

New PDFs

Believe it or not, the Revu app is capable of creating a new, blank PDF file. With a simple touch of the New PDF icon, as shown in Figure 8-15, the user will be prompted for a title. Entering a name and choosing a storage location will create a single-page, blank PDF. From here, the user may use all the tools within the app to generate a legitimate document.

Field Templates

Though new PDFs are possible, more often than not, the authors have seen field users start from some sort of template. Two specific examples are shared in the case studies of this chapter, but the examples are almost infinite.

A few of those possibilities are RFIs, Daily Inspection Reports, Quality Control Reports, and Work Orders. Frequently these templates are PDF forms that have prepopulated fields where users can enter the requested information. Templates are a great way to make sure key information or tasks aren't incidentally forgotten, whether that's part of the work to be done or pieces of data to be captured.

A simple Request For Information (RFI) form is shown in use on the iPad in Figure 8-16. This form can be created ahead of time in the office setting and then used repeatedly in the field.

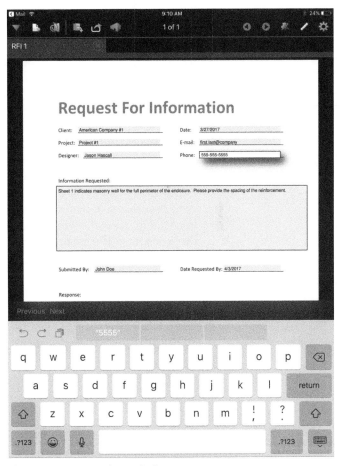

Figure 8-16: RFI Template on iPad

As shown, the forms are easy to complete on the iPad, but the field user can also add supplemental content and information like photos, attachments, additional pages, voice notes, and more. These items that couldn't have been done without returning to the job trailer can now be completed on the spot, making it much more likely to be done thoroughly and correctly. As an added bonus, the user will see in Chapter 9 that the data from multiple forms can be extracted into a table with one simple process.

Consistency Is King

From the authors' own experience, templates can be incredibly valuable for field inspectors. From welding inspections to paint inspections to general construction inspection, a templated form helps each inspector to remember what data he or she should be recording.

Many times these inspections are dictated by building codes or design specifications, but those documents can be long and cumbersome, making it difficult to recall specifics of the exact task at hand. For example, does a fillet weld on a tainter gate require magnetic particle testing or dye penetrant testing? The answer is, "It depends." Inspection templates can help with that by answering those questions by providing form field entries that indicate the correct procedures.

The pre-populated form blanks also help remind the inspector what data he or she should be recording and can also help maintain consistent formatting such as significant digits, units, and labels. Though a human can easily recognize and group "partly cloudy," "partially cloudy," "partly sunny," and "spotty clouds," the computer will identify four distinct groups unless the human informs the computer that the groups are the same. Drop-down menus inside of form fields completely eliminate that. They also remind the inspector to report the weather conditions even though he or she will be focused on the important task at hand.

As-Built Drawings

Another great application of iPads is "as-built" drawings, the documents used to convey any changes that happened in the field where the actual construction differed from the design. Traditionally, the as-built changes were marked in red pen on the hard-copy versions of the construction documents. With Revu, not only can the user mark the changes electronically, he or she can add photos, utilize screen shots from other sheets in the drawing set, and hyperlink to any relevant information such as RFIs, vendor websites, or other sheets in the drawing set.

And of course, with the latest cloud services, the as-built drawings can be stored in a location providing instantaneous access by all parties of the project team. Pre-iPad, the office team may have gotten lucky and received as-built updates every third week. In other cases they may only receive as-built drawings at the completion of the project. In either case, the lag is significant and may be detrimental to the construction process. Revu and iPads can completely eliminate that lag.

Accounts

In the authors' opinion, one of the trickier aspects for the user of the iPad app is knowing where he or she is storing the information. Frankly, Studio is so well integrated that it's easy for a user to save a file to a Studio Project when he or she thought it had been saved it to the iPad's local storage. A few screen shots should help clarify the interface for the readers.

Figure 8-17 shows the screen that appears when touching the file cabinet icon at the top of the Revu window. The first thing the reader should note is the house icon at the top, indicating that the root storage location is currently being displayed. As such, the reader can see the list of storage locations currently accessible by the app. In this specific case, the list includes the Documents folder, which is the iPad's local storage, and a number of Studio Projects that are cloud-based storage locations. In this example some names and Studio IDs have been concealed to protect the content within.

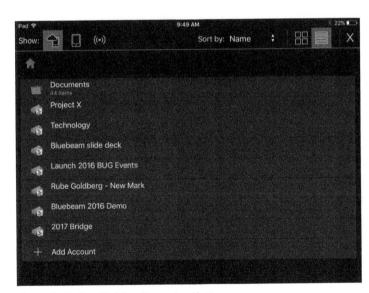

Figure 8-17: Revu iPad App Storage Locations

By touching the Documents line, the user moves to the Documents folder, listing all the subfolders and files currently stored on the iPad. Two things to note in Figure 8-18 are the "Documents" title adjacent to the house icon, indicating the location of the file list, and the iPad icons adjacent to each file indicating the file is currently stored on the iPad's local memory.

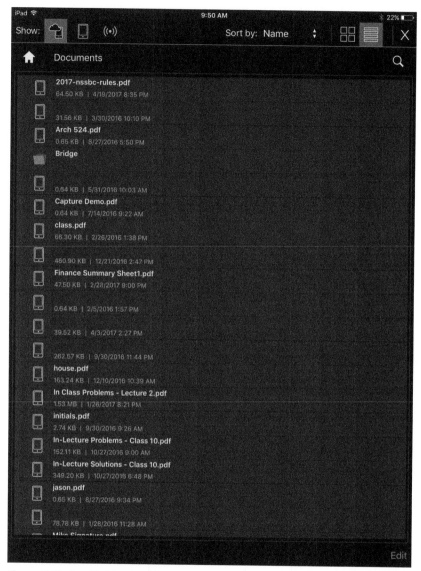

Figure 8-18: Revu iPad App Local Storage

Touching a Studio Project instead leads the user to the contents of that project. As shown in Figure 8-19, the location now indicates the name of the Studio Project and the icon adjacent to the file is a cloud, indicating the particular file is in the cloud, not locally stored or cached to the iPad. The user can also see the padlock icon on the right side, showing this file is currently checked in to the Studio Project.

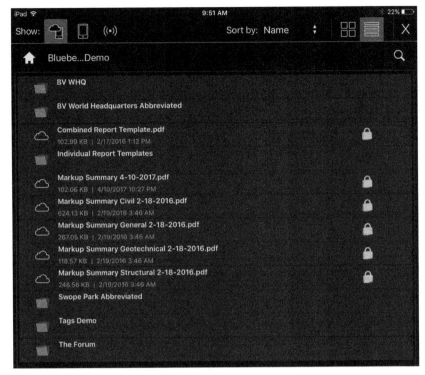

Figure 8-19: Revu iPad App Studio Project Storage

Bluebeam for iPad also integrates with other cloud storage platforms such as Box, Dropbox, and Microsoft SharePoint. Any of these integrations can be added from the Document Manager home screen shown in Figure 8-17 by touching Add Account at the bottom of the list. A short video located at www.bluebeam.com/solutions/revu-ipad provides a great demonstration on how to add other cloud accounts.

With a little practice, understanding and navigating the storage locations becomes a lot clearer.

CASE STUDY

Meet Greg Martin, Senior Manager, Operational Excellence, the Weitz Company

After Greg Martin attended the Bluebeam eXtreme Conference in 2015, he returned to his Phoenix office with a big idea: a simple project dashboard.

Greg's previous employer had utilized dashboards well, but they were fancy dashboards. They were dashboards that were easy for users, but required web developers, Adobe Dreamweaver, and lots of complicated programming. What Greg saw at eXtreme was nothing like that. Professionals with no experience programming a computer were making icons and quickly linking them to documents and folder locations, building a dashboard in a matter of hours, not the weeks Greg was used to.

Greg first instituted the idea on a student housing project at Texas A&M University in College Station. The project included 16 different buildings over approximately 70 acres, with each building being a little different than the others. At its peak, the project had over 1500 professionals working at the same time, meaning communication was going to be both important and challenging. "It just made sense to me that Weitz needed a visual way to communicate and provide access to project information," said Greg. The dashboard could be a perfect fit.

The Weitz team set off on a course to design their own dashboard. Almost immediately, they knew they wanted a map-based dashboard rather than a building name or number type dashboard.

Greg noted, "An employee who walks on the jobsite the first day is going to be thrown multiple nomenclatures of all the buildings and areas. How is he or she going to know this is Building 14 or Building B? Instead, why don't we show him or her that same information on a map?"

So the team started with that. They created a visual dashboard that was a map of the site because everyone on site could point to the building they were standing in on a map, but they couldn't always tell you the correct name. It was a clear way to communicate and navigate to a building's associated plans.

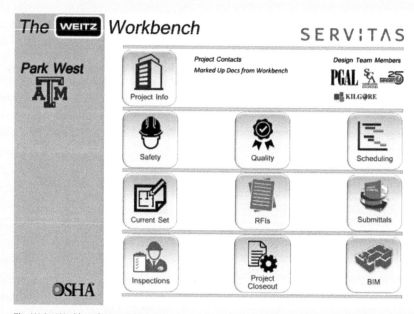

The Weitz Workbench Home Page

The WEITZ Workbench

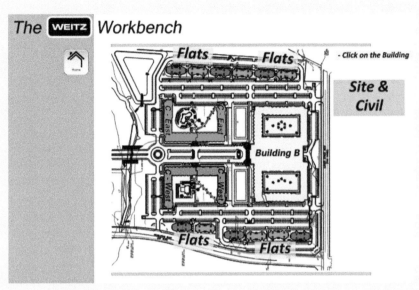

The Weitz Workbench Site Plan

Weitz went through about five or six versions of the dashboard before they settled on the final answer. Greg said that most of the time was spent playing with the look of the dashboard, the aesthetics. Matching the Weitz brand was important, but also finding a simple, easy to understand look for the users was the overall goal.

Greg's team also incorporated a lot of feedback from the field. The team always went back to a single question, do the workers in the field have access to what they need and what they want? Adding and removing things based on their feedback, the team reached a solution that worked for everyone.

"In the end, it took longer to come to an agreement about what the dashboard looked like and what links it contained than it took to actually build the final product." Greg said, and with that, the Weitz dashboard was born.

The next step for Greg was putting this dashboard to use. On this project, Weitz would use Box as the central hub for all project-related documents and information, with each instance of Bluebeam pointing to that central Box location. With a project of this size, all project team members were located onsite and all project information was managed onsite. The team assigned one project engineer to be responsible for making necessary adjustments, modifications, and updates to the dashboard and the associated documents.

Every professional on the Weitz team was given an iPad with a copy of Revu and each subcontractor on site was shown how to download and use a copy of Vu, Bluebeam's free mobile PDF reader software. With Vu, subcontractors were only capable of read privileges, so even outside of the Box permission configurations; there were no concerns of subcontractors inadvertently modifying the project documents. They would have on-demand access to the project information they need at the location they need it.

Greg and the Weitz team began the project with one very large subcontractor training session and dashboard setup. It was part of their normal "Site Onboarding" where they cover topics ranging from safety to schedule to site logistics. The dashboard training was a small piece of that and included both iPad and "Weitz WorkBench" training. The Weitz WorkBench is a digital plan table setup that includes two flat-screen TVs, which allow anyone on the project to access and view critical project information, such as plans, specifications, and the coordinated project BIM.

During the training, the trainers simply showed the attendees how to access and use the dashboard on all the devices. "The training was short," said Greg, "It's pretty intuitive to touch a building in order to access the plans and documents." The team concluded by configuring each and every professional's iPad to have the dashboard pinned in his or her recent documents so it was always easy to find and access.

From that point on, the team was let loose. If a professional needed a project document, he or she went to the dashboard. If someone wanted to check for updated versions, again he or she looked to the dashboard. All project information was accessed through the dashboard, leaving no need to navigate any type of folder structure. Weitz still managed all project documentation following standard electronic document management practices, but for the workers putting the work in place, they did not need to know exactly which folder each document was in; they just needed to know that they could access it from the dashboard.

Perhaps the best part was that the formal training ended there. Once a professional knew where to find the dashboard and how to use it, there was nothing else to learn. When new professionals came on board, they were simply peer taught. They found a colleague who quickly showed them the ropes and off they went.

Greg noted that the subcontractors really liked the simplicity of the dashboard, specifically noting the benefits of the visual interface and the map. Everyone liked the map. "I click on and it takes me there. It doesn't get easier than that," noted one subcontractor.

The dashboard did evolve during the project. Several links were added that weren't there originally. One subcontractor requested a quick link to a schedule in Bluebeam so multiple parties could coordinate logistics. The result created a way for one sub to coordinate with another simply by clicking a hyperlink. There was no longer a need to have a printed schedule in hand. No worry about it being current or losing it in a truck. The most current version would always be accessible from the dashboard, just like everything else.

Overall, the dashboard was an incredible success. The whole team was happy and Weitz has since implemented it on new projects they've started. That said, Weitz did identify some tips and hardships for others developing dashboards:

1. Set a rule that no document can be deeper than four touches. If it takes more than four to open a document, the users get frustrated.

2. The hardest part was linking the Box account to Bluebeam on the iPads. Though the training team took care of that for the initial wave of professionals, it was a hardship

for professionals coming onboard midway through the project. They solved it by posting instructions on two 11 × 17 sheets of paper in each trailer. The instructions were so easy that anyone could do it, or at least find a tech-savvy colleague to help.

3. Even though users could still technically access the documents individually via the Box App, Box doesn't recognize relative path hyperlinks, and so there were times Weitz would have to explain that all the valuable links work when the documents are accessed on the mobile device via Bluebeam Vu and not Box.

4. Acquiring iPads was difficult for some subcontractors. Every professional has a phone, but Bluebeam Vu doesn't currently exist for iPhone or Android, so that wasn't an option. Such an app would enable an even larger adoption rate of the dashboard.

The Weitz Workbench Building Page

The Physical Weitz Workbench

In the end, Greg felt really good about the team's accomplishment. They were able to create something great with very little effort. The dashboard proved to be a win for everyone and continues to be a great tool for the Weitz Company and their partners.

Recognizing the value of the presenters at eXtreme and their influence on his solution, he went on to help Bluebeam form the Phoenix Bluebeam User Group (BUG), a locally based group of professionals that meets at a regular frequency to talk about all things Bluebeam.

Similar to the authors of the book, Greg and Weitz have found great value in the BUG groups. In fact, that's exactly where the authors met Greg for the first time, at a Kansas City BUG meeting, where Greg was sharing with others about the Weitz dashboard.

Conclusion

Being mobile is no doubt important for any AEC firm, and in today's world it's completely possible. Bluebeam's iPad app can no doubt play a role in most firms' field workflows. Overall, the iPad app is a great tool for the mobile user. Although not as powerful as its PC-based older brother, the app has a lot to offer, and at just $10 its value can't be beat. Users interested in learning more can check out a list of video tutorials specific to the iPad at http://support.bluebeam.com/ipad/#wp-video-lightbox/10/.

Chapter 9
Go Digital, Document Assembly

For some people, the benefits of going digital are numerous and obvious; and for others, that statement holds true for staying nondigital. Falling into the first category, the authors of this book have discovered that Bluebeam is one software that makes going digital possible. They've taken that to heart and have personally operated entirely paperless in the work environment for over two years.

Although a common theme of this book is how Bluebeam enables digital workflows, this chapter will focus specifically on assembling documents and making them navigable and portable. Specific topics include:

- Forms
    - Creation Tools
    - Automatic Form Creation
    - Data Merge
- Functions
    - Optical Character Recognition
    - Reduce File Size
    - Create and Repair
- Batch Functions
    - General
    - Link
    - Summary
- Document Security

PDF Forms

PDF forms are very handy when the user is looking for specific types of responses from other users or him- or herself. They work extremely well for any type of document that is frequently repeated such as a Request for Information (RFI), a Daily Inspection Report, or a Survey.

Forms, at their basic level, provide input areas where a user may enter requested information. The creator of the form has the ability to restrict the type of input (text versus number), length/size of the input (number of characters), or even the discrete, specific entries (yes/no, red/blue/yellow, etc.).

Forms are beneficial because they can remind a user of the information that needs to be captured and control the information such that it is consistent each time the form is used. In the case of a Daily Inspection Report, a form may remind the inspector to document the outdoor temperature and weather conditions, a factor he or she may frequently forget. Further, that form may indicate the temperature should be recorded in degrees Fahrenheit as opposed to degrees Celsius such that two different inspectors record results in the same units.

The following paragraphs will discuss the various methods of form creation and data extraction within Revu.

Creation Tools

The most basic way to create a form is to utilize the creation tools found within the Forms Tab shown in Figure 9-1. The individual tools may be identified by hovering with the mouse above the tool and most are self-explanatory by their title. For instance, the Text Box tool creates a text entry box; the dropdown tool creates a dropdown selection menu.

Figure 9-1: Forms Tab and Tools

Each tool can be used in a manner very similar to the standard markup tools; simply click the tool and draw a box in the document where the desired form should be located. In the example shown in Figure 9-2, text boxes have been added to the top six entries in the generic RFI form, with one of the forms selected in Edit mode.

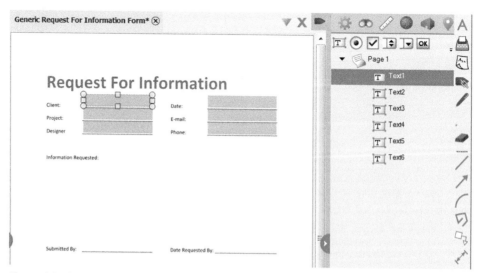

Figure 9-2: Generic Form with Text Boxes (Edit Mode)

Before moving on, it's important to differentiate Edit Mode from Fill Mode. In Edit Mode, the user is able to adjust the size, location, and properties of a form. In Fill Mode, the user can only fill the form as desired by the form's creator. Edit Mode is entered simply by selecting any one of the forms in the list on the right. When selected, the specific form will display yellow sizing handles on the corners and sides and all the form fields in the document will become a darker shade of blue. At this stage, the user can select any form field to edit in the list on the right or directly within the document. To return to Fill Mode, simply hit the Esc button, at which point clicking on a form field in the document will allow the user to enter a meaningful input value.

To edit the characteristics of a given form field, the user can utilize the Properties tab, similar to a standard markup. To do so, simply select a form field in the list on the right to enter Edit Mode and then click on the Properties icon to open the Properties Tab. The available options within the Properties Tab will be different for each type of form field, but the various settings are fairly intuitive.

As an example, Figure 9-3 shows the same form as before with an additional text box form field under the "Information Requested" heading. Because this form field is clearly intended to facilitate paragraphs of information, it's important that the form creator selects the Multi-line option in the Properties Tab on the right. Without doing so, the entry will be limited to displaying one line of text only. The reader should also note the variety of other options available to edit the properties of the field, including Font, Font Size, Text Color, Character Limit, Alignment, and more.

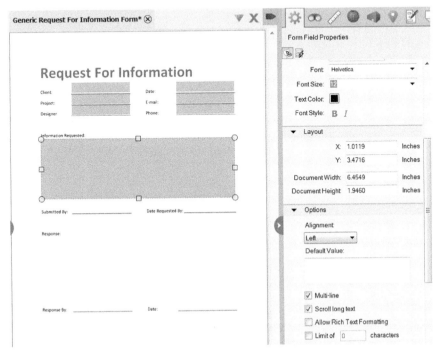

Figure 9-3: Text Box Properties

For additional assistance with form fields, readers will find a short training video available for free on Bluebeam's website located at: http://bluebeam.com/us/bluebeam-university/training-materials/pdf-forms.asp.

Helpful Tricks

The form creator will be happy to discover that all the snap, alignment, and copy tools associated with markups also work for forms. Thereby, if the creator has two text boxes he or she would like to make the identical size, both lining up on the left, he or she can do so by selecting them both and clicking the Align Width/Length button(s) followed by the Align Left button, just as one would do for the rectangle or the cloud tool.

The distribute tools also function, so radio buttons that are not quite the same spacing can be easily adjusted to please even the most persnickety user. Figure 9-4 shows before-and-after states for both tools.

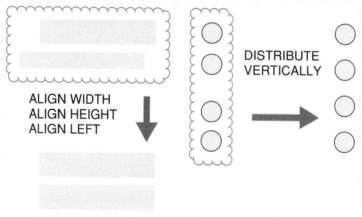

Figure 9-4: Align and Distribute Tools for Form Fields

Tricky Naming

A quick second tip is that no two form fields may be named identically unless the contents of the form field are intended to be identical. This is a requirement of form fields in all PDF software and can become an issue when similar documents such as RFIs or Inspection Forms are compiled into one larger document. Often the existing form fields in each document are named the same as the base template, and combining multiple documents into one can cause issues with the contents of the field. As such, the reader should use caution when combining documents with form fields.

Automatic Form Creation

With the release of Revu 2017, Bluebeam introduced Automatic Form Creation, a tool that will take a template scanned from a hard copy or created in Microsoft Word or another native form template creation software and add form fields based on the structure of the template. A quick video located at www.bluebeam.com/solutions/revu2017 highlights the feature and capabilities of the new creation tool.

Historically, creating PDF forms could be a time-consuming task in Revu. Though often worth the effort by generating time savings in the use of the form on the back end, the new tool makes form creation a snap. Before diving into the details of automatic form creation, it's important to note that the new feature is found in the eXtreme edition only.

The first step in the process is to create a form template by scanning a hard copy or creating a background layer with Word or other native editing software. This is the content that may not be edited by the user, content that stays permanently and indicates to the user what values should be input into the fields. For instance, in the RFI form shown above, the title, values such as Client, Project, and Designer, along with their corresponding blank line are part of the native template.

As an aside, the template could also be created directly in Revu. However, the reader will likely find PDF less capable in the way of word processing than other native tools. If Revu is chosen as the creation tool, simply flatten the markups onto the PDF content layer to generate a working template.

Once the template has been created, select "Auto Create" from the Forms menu and Revu will go directly to work. If Revu recognized the template as a scanned PDF, the user may first be prompted to run Optical Character Recognition, or OCR, which will be discussed in detail later in this chapter. If the form template has been created natively with searchable text, Bluebeam will process the form and when complete, the user will see the newly added forms appear in the document, as shown in Figure 9-5.

The reader should note ten additional form fields added to the PDF, that each of those fields is a text box, and that three of those fields have been named according to the input description. So why have the other seven fields not been named, and why have the Information Requested and Response boxes not been created? The next paragraphs will explore the nuances of the tool.

First, let's consider the naming of the text boxes. In Figure 9-5, note the proximity of the entry line to the line's title. The Submitted By, Date Requested By, and Response By titles are significantly closer to their corresponding lines than the others such as Date. After returning to the native document and reducing the spacing, the reader will see in Figure 9-6 that the text boxes have now been named appropriately.

Second, let's consider the missing Information Requested and Response boxes. The reader will note that the native file contained no visible boxes or other defined space for these entries. Without the defined area, Revu is uncertain whether the text is meant to be an entry or simply a segment of text for the reader. It assumes that an input field is not required. Adding a rectangular input box in the native document will clearly inform Revu that a box is desired and the Auto Create function will work as expected, also shown in Figure 9-6.

Figure 9-6 also indicates that in addition to adding the text box, Bluebeam has correctly named the larger text box "Information Requested" based on the proximity of the label.

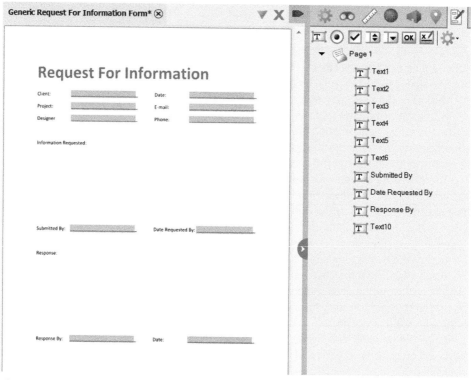

Figure 9-5: Auto Create Form Results

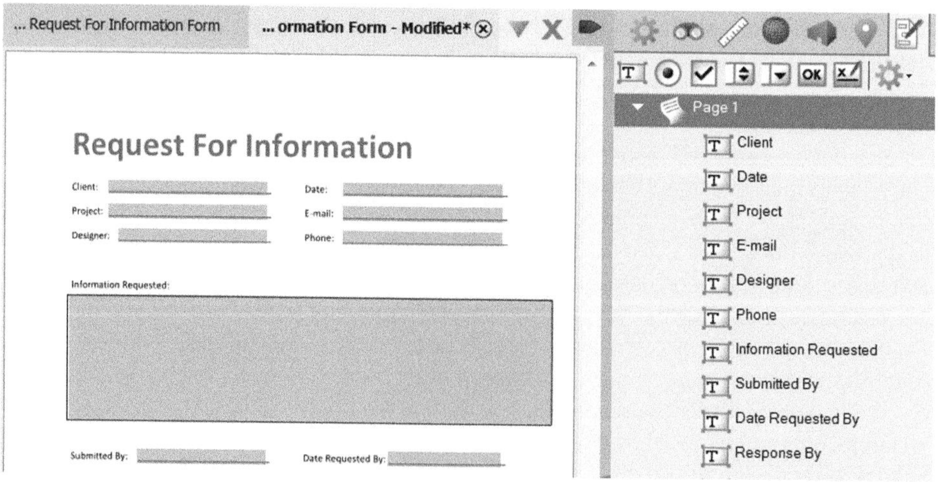

Figure 9-6: Modified Auto Create Form Results

And finally, a quick investigation of the properties of that text box will show the user that, recognizing the size of the field, Revu has automatically checked the Multi-line option to accommodate more than the default single line of text, as shown in Figure 9-7.

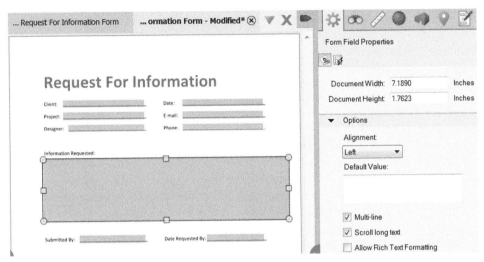

Figure 9-7: Non-Default Text Box Properties

As shown in the highlight video and noted in the multi-line text box, the creation algorithm does its best to intuitively format the fields. If Revu finds a line labeled Signature, it will create a digital signature field. If Revu finds a pair of empty parenthesis or a small box, it will create a checkmark field. If Revu creates a box labeled Date, it will be formatted for date entry. All three of these offer logical results that will generally save the user significant amounts of time.

Don't Be Fooled

To date, the Auto Create form field creation function has been functioning very well for the authors, creating most fields as would be expected. The reader should recognize, however, that the tool is certainly not foolproof.

A quick example is the creation of the multi-line text box described in the paragraph above. On the initial modification the line weight was reduced to 0.5 pt to lighten the appearance. Unfortunately, Revu didn't find the box at all and the line weight had to be increased to 1.0 pt.

As a user of Bluebeam, the reader should recognize that auto creating form fields require a little finesse. If the first attempt is unsuccessful, try again with some modifications. It may take a few attempts to solve, but it will save hours of time in the long run after the user develops a knack for creating the fields.

Data Merge

The final form field function that will be specifically covered by this chapter is the data export, import, and merge capabilities. Often a form is created as a way to acquire consistent responses to a set number of inquiries. A form could even be a survey in some cases.

In those cases, it's significantly more beneficial to have the data summarized in a table than dispersed in multiple single file forms. In a table, the data can be utilized to quickly identify trends or even conduct higher-order analysis.

Once a form has been created, distributed, and completed by a number of users, or a single user over a number of events, Revu can be utilized to quickly compile the responses into a spreadsheet. Under the Form menu, the reader will discover the Data menu ≣ᵣ, and under the Data menu, three options, Import Data, Export Data, and Merge Data. Export and import function as expected, both a simple way to move form field data for a single file. Merge Data allows the user to combine data from multiple files.

To utilize the function, simply select Merge Data. In the Merge Form Data window, shown in Figure 9-8, add the appropriate completed PDF form files and select OK. Revu will prompt the user for a file name and an export file type, csv or xml. The data will be extracted from each form file and compiled into one database, as shown in Figure 9-9.

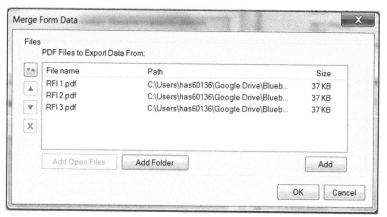

Figure 9-8: Merge Form Data Window

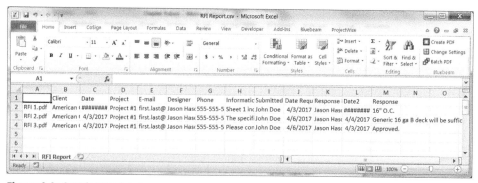

Figure 9-9: Sample Form Data Summary

Functions

Revu also includes a handful of functions to improve the usability of documents, three of which will be covered in this chapter.

- Optical Character Recognition (OCR)
- Reduce File Size
- Create and Repair

Each function has a very specific purpose and each has been beneficial to the authors at one time or another in their careers. Use of the three functions isn't nearly as common as the markup features, but when they are needed, they are very useful.

Optical Character Recognition (OCR)

In simple terms, Optical Character Recognition, or OCR, is the ability of the computer to recognize fonts that are not selectable and convert the characters to selectable and searchable fonts.

With today's technology, documents, drawings, and other files that are created in an electronic format should be created with selectable fonts. Therefore, any PDF files that originate in software like Word, Autodesk Revit, or Autodesk AutoCAD should contain selectable font. If they do not, the issue is likely a simple matter of changing some settings in the process of creating the PDF file. OCR should never be the "best practice" but rather a resource to use when needed.

Let's take a look at two examples where OCR might be required:

1. Scanned Documents: Sometimes designers and engineers only have access to hardcopy documents and drawings, or scanned images of those hard copies. This may be because the drawings are historic, created before electronic methods of archiving prevailed. This may be because the owner of the documents and drawings has only provided them in hard copy or scanned format.

2. Received Documents: Sometimes documents or drawings are provided electronically, but with nonselectable text. The provider may not have the technology to create selectable text, the provider may not want to take the time to re-create the files with searchable text, or the provider may be legally prevented from providing selectable text. In other cases, the document provider may have provided nonselectable text by pure accident, but the client relationship is such that it isn't appropriate to request a resubmittal.

Certainly other cases exist, but these are two good examples where OCR can greatly improve the usability and functionality of the files. As discussed throughout this text, many of the advanced features that users have come to love require selectable text. From simple searching to batch linking, which will be discussed later in this chapter, selectable text expands the functionality of a PDF.

Running OCR

Before running the OCR process, the user should open the file of interest. To find out if the text within a file is selectable, change the cursor to the "Select Text" cursor and try to highlight a segment of text. As shown in Figure 9-10, the word FLOW cannot be highlighted, indicating it is not selectable, and OCR must be completed before the higher-order functions may be used. If the text was selectable, the user would see the actual text become highlighted. When the user is ready to run the OCR function, it can be found under the Document menu. Clicking the icon will launch the window shown in Figure 9-11.

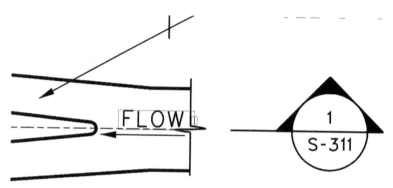

Figure 9-10: Non-Selectable Text

Figure 9-11: OCR Window

The window contains a number of options. By default, the OCR is optimized for draw-ings. Choosing an alternate Document Type will change the function algorithm to perform better on different features anticipated within the specified type. Though a correct Type selection will help the OCR results, the function will still process on any file, independent of the specific Type chosen.

Continuing with the default settings and clicking OK will cause the OCR module to run, identifying all the nonselectable text in the selected page range and converting it to select-able text. As shown in Figure 9-12, after the OCR, the word FLOW can now be selected with a highlight.

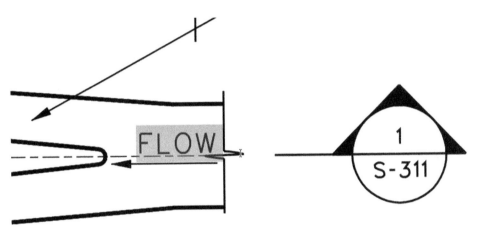

Figure 9-12: Selectable Text

Settings and Advanced OCR Capabilities

One of the great advanced capabilities of Revu for drawings is the ability to detect vertical text. Often on drawings, the title block and detail callouts contain vertical text. Revu is capa-ble of identifying and converting that text also. Without vertical text detection capabilities, the Batch Link function would not hyperlink any of the vertical section cuts.

A second feature to be aware of is the option to Skip Vector Pages. When a document contains a mix of pages with selectable text and nonselectable text, skipping pages that already have selectable text, vector pages, will save time in the OCR process.

The third feature to note is the availability of non-English Recognition Languages. Simply click Add at the top of the OCR window to select one of 13 alternate or additional languages.

Finally, the last option the author would like to specifically point out is the Page Chunk Size. This is the ability to change the quantity of pages sent to the OCR engine at one time. Sending more pages at a time may reduce the time it takes to run the overall OCR process, but each incremental page in the OCR engine will occupy a larger percentage of the com-puter's processing power.

OCR Imperfections

At the beginning of this section, it was noted that OCR should never be considered the "best practice." There are two simple reasons for this: 1) creating selectable text PDF files directly from the native files is faster, and 2) creating selectable text PDF files directly from the native files is more accurate.

The truth is that very few OCR runs are perfect for 100 percent of the text. The lower-case letter "L" may be mistaken for a "1" or an upper-case letter "I." Two subsequent spaces may be mistaken for one space or vice-versa. Spaces might appear where they didn't exist in the native document. Any number of problems may occur that result in selectable text that doesn't perfectly match the native text.

Though the human mind can easily look beyond these subtle differences and understand the intent, syntax is important to the machine. Searching for "SOIL" doesn't return a hit on "SO1L" where the letter "I" was replaced with the number "1." The same is true for some of the higher-order functions like Batch Link, which will be discussed later in this chapter. With Batch Link, exact matches are imperative.

Therefore, although OCR is a helpful tool, it should rarely be the go-to option. Instead, attempt to create PDFs with selectable text directly from the native applications.

Reduce File Size

Often times PDF files contain a significant number of images. In those cases, file sizes can become extremely large. Lucky for Revu users, the Reduce File Size function is capable of reprocessing PDFs into more reasonable files sizes. In fact, the authors have seen reductions in file sizes as high as 80 percent.

The function is simple to use and can be found under the Document—Process menu. Clicking Reduce File Size launches a brief status bar window, indicating Bluebeam is analyzing the document to determine the level of file size reduction possible and estimate the new file size. In general, the more images in the file, the more reduction that's possible.

After the brief analysis window, the user will receive the Reduce File Size window, as shown in Figure 9-13. Here the user is presented with the estimated file size and the percent reduction from the original. He or she also has the option to select the balance between Quality and Compression. The farther toward Compression on the right, the greater the file reduction, but the more grainy the images may appear. Moving the slider to the left, the image quality is maintained, but the reduction in file size is reduced. By default, Revu selects a balance of the two.

The careful observer has also noticed the Custom option with an Edit button to the right. Here the user can be very specific with the modifications being made to the file, but this highly advanced capability will not be covered by this book.

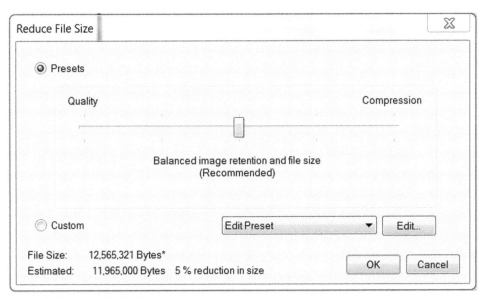

Figure 9-13: Reduce File Size Window

EXPERT TIP

Review after Reducing

As noted, file size reductions can be extreme, above 80 percent. However, those reductions don't come without risk. On occasion the image reprocessing doesn't work as desired. The authors have seen instances where images were replaced with white or black rectangles, instances where images were scaled or distorted, and instances where specific pages result in a display error.

These cases are very rare, but it's still a good idea to scroll through the full document after running Reduce File Size. Few things are more embarrassing than submitting a drawing or document to a client with a large open space where an image should be. A quick review takes only minutes and assures no unexpected behavior.

Create and Repair

The final function covered in this section is the Create function, which was introduced in chapter 2. Although the title is Create and Repair, the function is only titled Create. It can be found under the File menu with the icon ▣. The section title includes Repair because the function is capable of doing exactly that.

The authors considered covering Create with an Expert Tip, but the function is so valu-able that it deserved its own section. Create generates a PDF file utilizing the most stable, pristine, and thorough algorithm possible. It is the process least likely to generate bugs or errors in the resulting PDF. It is also the slowest creation process available within Revu. For that reason, it's not the algorithm typically used when creating PDFs, but it is the algorithm capable of repairing bugs that might occur with the default process.

If the user finds a PDF file that is not behaving as expected, Create may be a resolution. For instance, perhaps the file isn't displaying correctly, or maybe scrolling to page 56 throws an error message that locks up Revu, or possibly a certain file tends to crash frequently. All of these are instances where recreating the file utilizing the stable Create process may resolve the problem.

To use the function to repair a PDF file, simply click File—Create—From File. This will bring up the standard Windows Explorer window, giving the user the ability to select the file of interest. Once selected, a second Save As window will appear, allowing the user to specify a file name and folder location. Upon clicking Save, a progress bar will appear indicating the status of the creation, and once complete the file will open in the main Revu window.

The Create button also has several other functions worth noting, as shown in Figure 9-14. The From Multiple Files option allows the user two options, to re-create multiple files and maintain multiple files, or to re-create multiple files and combine them to one. From Scanner or Camera creates a PDF from an attached external device. Layered PDF creates a single PDF from multiple PDF files with each individual file existing on its own individual layer. Finally, PDF Package groups multiple PDF files into one Package that acts like a container for the individual files, somewhat like a zip folder.

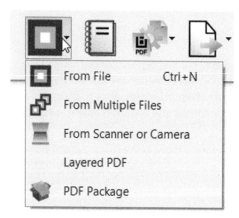

Figure 9-14: Create Menu

Batch Functions

Batch functionality within Bluebeam is the ability to run a process or function on multiple files at one time, with a single action by the user. The batch menu has been a part of Revu for many years and continues to grow with every release.

The time savings realized by the user are almost immeasurable due to the potentially unlimited number of files. For a user dealing with only a handful of files, the batch processes may not offer much time savings, but for the user with hundreds or thousands of files, the implications are enormous.

Batch processing frees the user to do other tasks while the computer is repeatedly processing one file at a time. Batch processing reduces errors resulting from a user repeating a mundane task over and over. Batch processing improves consistency of drawings and documents by applying the same rules and algorithms to all the files in the batch. Batch processes make the user more efficient.

General

The Batch menu can be identified by the icon 🐝 that can be found under the File menu. The full list of batch functions available in Revu 2017 is shown in Figure 9-15. Most of the functions in the menu are standard Bluebeam functions that have been expanded to operate on more than one file. In each case, the user will be prompted to select the files he or she wishes to operate on before navigating through all the options of a given function. In a few rare cases such as Batch Link, there is no corresponding single sheet process because the operation isn't applicable to one sheet. Below is a brief summary of the available options, each capable of operating on more than one file.

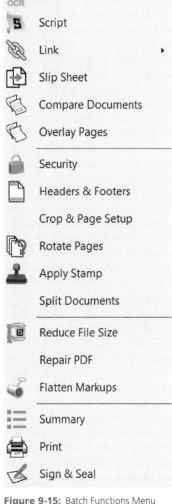

- OCR: Previously discussed in this chapter, OCR is the ability to transform non-searchable text in a file into searchable text within that same file.

- Script: The ability to utilize Java Script functions to perform operations inside PDF files.

- Link: Discussed in detail later in this section, it is the capability to automatically find and hyperlink references within a set of files.

- Slip Sheet: Previously discussed in chapter 5, the ability to replace or supplement old versions of a sheet with a new version.

- Compare Documents: The ability to identify, cloud, and label differences between similar documents.

- Overlay Pages: The ability to identify differences between similar documents by stacking files on top of one another, similar to a light table.

- Security: The ability to apply security restrictions to a file.

- Headers & Footers: Discussed in chapter 2, the ability to create and modify PDF headers and footers.

- Crop & Page Setup: The ability to adjust parameters associated with the page; size, orientation, scale, and the like.

- Rotate Pages: The ability to rotate the orientation of the pages within a PDF file.

- Apply Stamp: The ability to apply a standard stamp to a PDF drawing or document.

- Split Document: The ability to subdivide a PDF document based on a certain criteria such as page count, file size, or bookmarks.

- Reduce File Size: The ability to reprocess images in a manner that reduces the size of a PDF file.

- Repair PDF: Previously discussed in this chapter, the ability to recreate a PDF file using a stable process to improve its performance and remove any "bugs."

Figure 9-15: Batch Functions Menu

- Flatten Markups: As described in chapter 3, the ability to move markups from the markup layer to the PDF content layer.
- Summary: Discussed in the subsequent paragraphs, it is the ability to summarize the markups within a file for sharing, exporting, or reporting.
- Print: The ability to physically print a file to hard copy or PDF.
- Sign & Seal: Specifically discussed in chapter 10, the ability to apply an engineering seal and corresponding digital signature to a document.

Link

Although it has now been around for many years, Batch Link is one of the features of Bluebeam that differentiated itself in the market. Batch Link makes a set of drawings infinitely more navigable in a matter of seconds.

For those readers not familiar with engineering drawings, each drawing typically contains section cuts and detail bubbles that reference information on other sheets, as shown in Figure 9-16. This section cut is found on Sheet S-801, but directs the viewer to Section A on Sheet S-803, as shown in Figure 9-17.

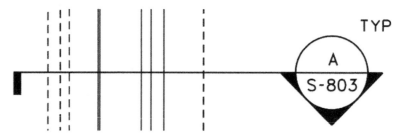

Figure 9-16: Sheet Reference Example

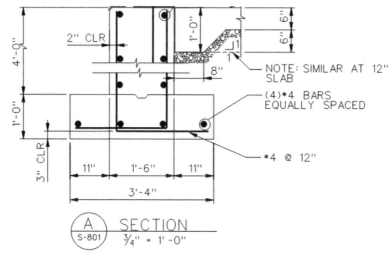

Figure 9-17: Reference Section

When using a typical hard-copy or electronic drawing set, the user has to identify the sheet reference in the section cut (S-803), and scroll or flip to that sheet in order to see the detail. In some cases, drawing sets may be hundreds of pages, and it's not uncommon that one reference may jump 10, 50, or 100 sheets. In hard-copy format, the user is able to grab multiple pages and quickly flip to the sheet of interest. In electronic format, the user had to scroll through the pages one by one, or recall the page count number of the referenced sheet. Both options were painstaking and nearly unmanageable. Truth told, this is the principle reason for resisting a shift to electronic documents. Users preferred to flip pages because the alternative was impossible.

Batch Link changed all that. By creating hyperlinks at each one of those drawing references, the arduous task of scrolling through hundreds of pages was reduced to clicking a single button. The ease of flipping pages was no longer an excuse for not going electronic.

New Batch Link

To run the Batch Link process, the user should click File > Batch > Link > New. Doing so will launch the Batch Link window, as shown in Figure 9-18.

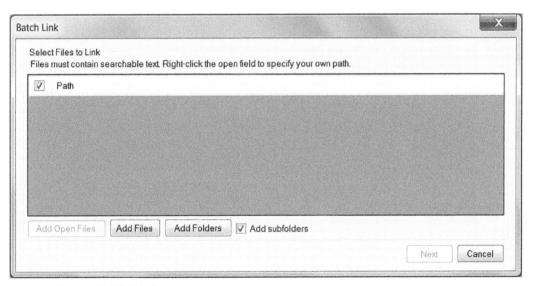

Figure 9-18: Batch Link File Window

Here the user may add a single multi-page file, multiple single page files, multiple multi-page files, a folder containing multiple files, or the files that are currently open in Bluebeam. In fact, any combination of the above will function in Batch Link. Once selected files are added to the list, the user may click Next.

Generating Search Terms

The second Batch Link Window, shown in Figure 9-19, is earnestly where all the magic happens. It's here where the user has the important job of generating search terms, which are the text strings the user wishes to hyperlink. In the sheet reference example from above, the desired hyperlink text string is the "S-803" reference in the section cut. The user would like to add a hyperlink to that text that will take him or her to sheet S-803 where he or she can find Section A. So, the desired search term is "S-803," or in more generic terms, the sheet number.

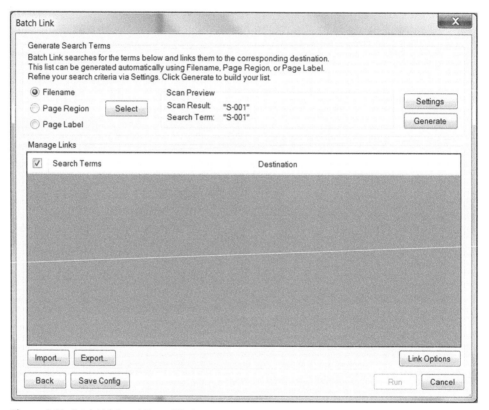

Figure 9-19: Batch Link Search Terms Window

In fact, the user wants to replicate this for every reference to any sheet in the drawing set, so really the user wants the list of Search Terms to match the list of sheet numbers. In line with that, the Destination of the hyperlinks created on each Search Term should be the page that the Search Term refers to, in the example above, the file and/or page number that corresponds to S-803.

Now that we've established what a Search Term is, let's consider the ways the user might create or "Generate" those terms. In the top left portion of the window, isolated in Figure 9-20, the user will find three options: Filename, Page Region, and Page Label.

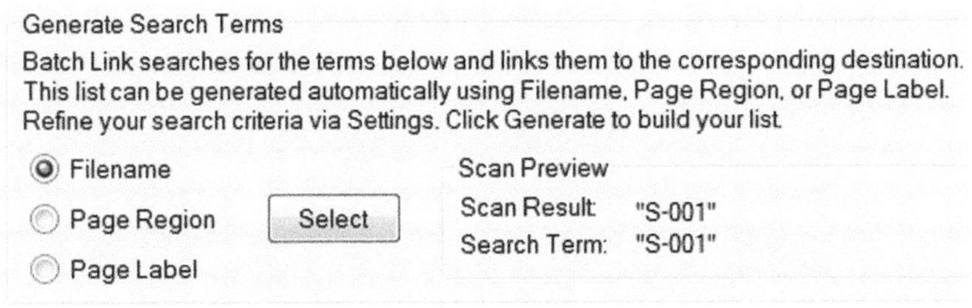

Generate Search Terms

Batch Link searches for the terms below and links them to the corresponding destination. This list can be generated automatically using Filename, Page Region, or Page Label. Refine your search criteria via Settings. Click Generate to build your list.

◉ Filename   Scan Preview

○ Page Region   [Select]   Scan Result: "S-001"
                             Search Term: "S-001"
○ Page Label

Figure 9-20: Search Term Options

The format of the files selected in the previous window will determine the method of generating search terms. If each file is a single page with filenames matching the sheet numbers, then Filename would be a perfect option. If all of the files selected contain page labels that match the sheet number (potentially created using the Automark feature), then the Page Label option would be a great choice. If neither of the above is true, the user should choose Page Region.

If the user selects the Page Region option, one more step is required. After clicking the Select button, Revu will launch one of the files and give the reader the opportunity to place a rectangular selection box around the sheet name in the file. Revu then remembers that location on the sheet and looks in the same place on every page to find a search term for every page in the selected files.

Smart Documents Equal Smart Results

Throughout this text, the authors have made the case to create the PDF documents well; vector content, selectable text, and consistent page sizing. The Batch Link feature is one big reason for that.

If it's not completely apparent to the reader already, Batch Link can save hundreds or thousands of hours on a project. In fact, it can make the difference between utilizing electronic documents and being forced to print hard copies. It is one of the most significant features available in Revu today.

Now, imagine if the files selected for Batch Linking did not have searchable text. Although search terms may have been generated, they will never be found within the group of selected documents because those documents do not contain searchable text. Batch Link would not create any hyperlinks and the user would once again be scrolling through hundreds of pages to find the one detail he or she is looking for.

Similarly, if the files are sized or oriented inconsistently, the Page Region option may not find a search term. Imagine if one specific page was smaller than all the rest by 2 inches in each direction. The page region selection window would likely land off the page in empty space, returning nothing.

It is advanced features like Batch Link where these subtle nuances make all the difference.

Once an option has been selected, click Generate and Revu will populate the list of Search Terms and Destinations. In the example from above, Filename has been selected and the results are shown in Figure 9-21. Note in the example that the Destinations are individual PDF files because the selected files are each one page long. After the hyperlinks are created, clicking one of them will launch the Destination file for the user.

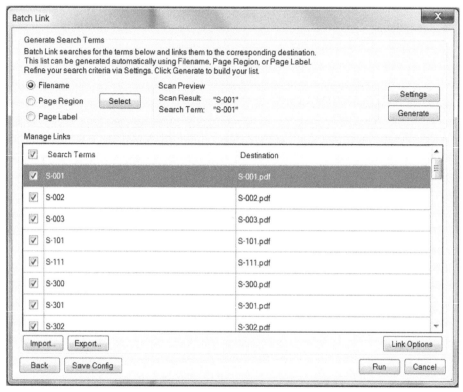

Figure 9-21: Search Term and Destination Results

It's important to recognize that it's very possible that more than one of these options will work and it's also very easy to test each out; simply click the Generate button and see if the results are as expected. If not, Cancel and start again.

Before moving on, there are a few other options important to point out. First, take a look at the window behind the Settings button, shown in Figure 9-22. This window conveniently allows the user to specify a filter on the search term generation algorithm. For instance, maybe all the file names begin with a project number that is not included in the reference callouts within the sheets. This feature would allow the user to remove that project number automatically. In the lower left corner of the search term window, the user will see the options to Import and Export the search terms. These capabilities may come in handy when a list of search terms exists in spreadsheet format. Finally, the Save Config option, also in the lower left, allows the user to save this search term generation setup for reuse, perhaps following a round of document revisions. It's a simple way to save time in the future.

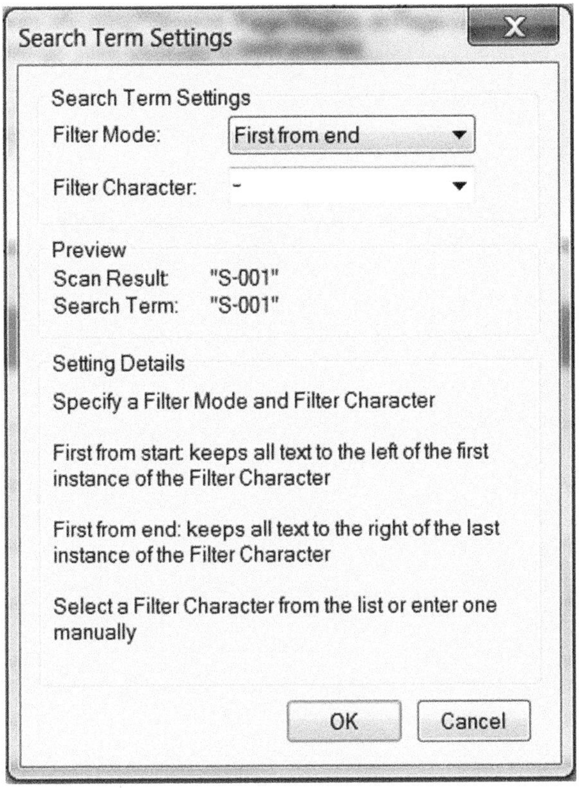

Search Term Settings   X

Search Term Settings

Filter Mode: First from end ▼

Filter Character: ~ ▼

Preview
Scan Result   "S-001"
Search Term: "S-001"

Setting Details

Specify a Filter Mode and Filter Character

First from start keeps all text to the left of the first instance of the Filter Character

First from end: keeps all text to the right of the last instance of the Filter Character

Select a Filter Character from the list or enter one manually

OK   Cancel

Figure 9-22: Search Term Settings Window

Hyperlink Options

Before clicking Run to move on, the user has one more important button to visit, Link Options. Though covered last, it contains the most important settings for the process. After clicking the button, the user will see the Link Options window shown in Figure 9-23.

At the top of the window, the user has two search options that may become very important for the results, depending on the exact drawing setup. Searching "Whole Words Only" may not work for users with prefixes or suffixes, but for others it will help reduce extraneous search hits. Similarly, the "Case Sensitivity" may help or hinder depending on the use case.

The lower portion of the window contains the options for the hyperlinks. The first choice is relative paths and is dependent on whether the location of the linked files will be moving. If they are all moving together, it makes sense to use relative paths. If they are moving independently or not moving, fixed paths are best. The second option is for those instances where a Batch Link is being rerun for updates to the drawing set or added sheets. In many cases the user would prefer to remove the old hyperlinks and replace them with the new, decluttering the drawings. The third option concerns the hyperlink's appearance. If the user wishes to have some type of appearance on the hyperlink, he or she should check the box

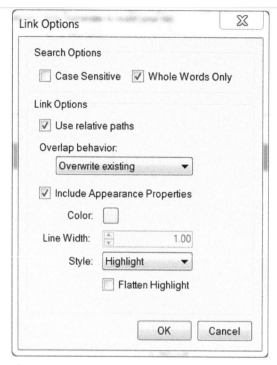

Figure 9-23: Link Options Window

and choose the appropriate settings. Two popular options are to highlight and underline. Finally, since the appearances of the hyperlinks are actually markups on the page, the user has the option to flatten those markups, removing the live content from the sheets.

EXPERT TIP

Remember the iPads

Often a user may consider transparent hyperlinks to keep a drawing looking clean and prevent clutter in the markups list. On PC, that's no problem. The mouse cursor changes to a pointing finger when crossing a hyperlink, indicating to the user that clicking in the location would take him or her to a new location. Additionally, the document can still easily be panned and zoomed without accidentally clicking any of the links because both functions can be accomplished without left-clicking the mouse.

The iPad, however, is a different story. There is no cursor and therefore no visual indication of transparent hyperlinks within a drawing. Because navigating, scrolling, and zooming require the use of one or two fingers, invisible hyperlinks may surprise the user and cause unexpected results.

Because the use of tablets is becoming so prevalent in the industry, the authors choose to utilize an appearance with every batch link.

Once the appropriate selections have been made, the user can return to the Batch Link window and click run. Revu will process through the sheets, create the hyperlinks, save the files, and when complete, return a Batch Link Summary window similar to that shown in Figure 9-24.

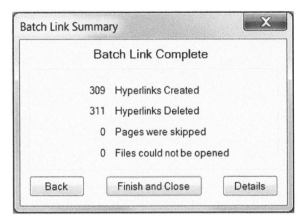

Figure 9-24: Batch Link Summary Window

In this example, the reader can see Revu is reporting that 309 hyperlinks were created during the process. At the same time, 311 hyperlinks were deleted. Recall that the over-write option was selected, so each time Revu found an existing hyperlink under a proposed hyperlink, it deleted the existing and created a new one. Additionally, the program noted no problems with opening files.

At this point the user should find hyperlinks at each sheet reference in any of the Batch Link files. An example is shown in Figure 9-25, where a reference indicates Detail A on Sheet S-527. The highlighted S-527 implies an existing hyperlink and indeed, clicking on the link will take the user to Sheet S-527.

Figure 9-25: Sheet Reference Hyperlink Sample

One caveat to note is that Bluebeam will not create self-referencing hyperlinks. In other words, if a reference found on Sheet S-527 referred to another detail on Sheet S-527, Revu would not create a hyperlink since the detail is on the same sheet as the reference and the user wouldn't be going to another sheet.

Meet Todd Parker, Senior Architect, Populous

It's not every day that a remarkable new tool just falls in your lap, but that's exactly what Todd Parker said happened to him with Bluebeam. "It just showed up on my machine. It sounds funny, but with over 150 software titles, it's bound to happen."

Todd Parker, Senior Architect

Kansas City–based Populous has around 300 professionals in the headquarters building with 14 other offices located worldwide. With notable projects such as Yankee Stadium and the London Olympics, the more-than-30-year-old architecture firm has a flair for public assembly spaces. Todd said, "We design everything from fairgrounds and equestrian centers to arenas and ballparks, but they all tie back to the places where people love to be together."

It was one of Todd's colleagues in the Kansas City office who actually pointed out Bluebeam to him, quickly making him aware of some of the benefits over other PDF tools. Todd, like many people, was surprised to discover that PDF was more than just reading documents and became curious enough to do more research outside the office. He recalled some deep investigation, including a lot of Googling. "I remember going to the 2015 release event at the Kansas City River Market. I went because I was interested to find out the differences between the Standard, CAD, and eXtreme versions," said Todd.

Fast forward a year from that event and Todd had already put Revu in place for a real project with an external partner, Turner Construction, but that's not where Populous started its work with Revu. "We started using Bluebeam internally first," said Todd. "It was mostly for internal coordination and markups and then grew into project reviews. As we learned more, we started to branch out and try more of the hefty commands like Batch Link."

Todd recalled that years ago most of the project review work was on paper, but today the processes require electronic submissions and the project reviews are completed in

Studio. He can't say if Bluebeam caused the shift to electronic reviews, but he knows it certainly didn't hurt. "Studio Sessions are great for reviews because we can get the comments while the review is still in progress. We don't need to wait and keep working in what could be the wrong direction."

Over the years since discovering Bluebeam, Todd and colleague Jason Gardner have helped convert their colleagues to Revu. Nicknamed the Bluebeam guys, the two men take the time to show other professionals how to use Revu and share the tricks they know.

Todd recalls that when he started the Anaheim Convention Center Betterment VII project, he didn't know they would be using Bluebeam. "It became apparent near the beginning of the project," he recalled. "Being a design-build project with Turner, who had prior experience with Bluebeam, it just made reasonable sense to share in Studio, so Populous suggested it and Turner was on board."

The project was an 850,000-square-foot expansion of the current Convention Center, including two levels of exhibition space, prefunction space, and a parking garage. In total, the project included over 17 design entities in addition to construction subcontractors, with Populous acting as the Architect of Record.

Anaheim Convention Center Betterment VII Rendering

With an aggressive schedule, Bluebeam Studio served as a tool to ensure that the contractor had the most up-to-date drawings as soon as they were available. "It's important to make sure the field has the latest information," said Todd. "The biggest thing for Populous was peace of mind that we have provided the most current documents to our client. If they used something else, it wasn't because we didn't get them the information."

Anaheim Convention Center Betterment VII Dashboard

Populous also uses Bluebeam for communication during construction, leveraging both Studio Sessions and Studio Projects. Todd explained, "As issues are raised, we commonly utilize Bluebeam to highlight areas in question. We can mark it up and begin the dialogue to develop the solution with the contractor. Working collaboratively, we finalize the approach and then document the solution through the RFI process using the Bluebeam PDF documents as supporting evidence attached to the RFI record."

Batch Link was a new feature Populous utilized on the Anaheim project, automatically hyperlinking the drawing set. But they've also been exploring ways to maximize the use of the tool to connect the whole construction document package, not just the drawings. Ideally they will be able to link drawings, specifications, schedules, and product data. "It streamlines the effort required to steer through these documents, which simplifies our efforts during the development of the package and assists the end users in their use of the documents as well," said Todd. "It's intuitive and easy to implement. Once people understand what you have done, they start seeing that and using it. The response has been very positive."

One step Populous is taking to enable this multi-document hyperlinking is the development of a standard abbreviation and nomenclature list. Instead of using a similar list from project to project, the list would be exactly the same, building consistency not only between different projects but also within a project since uncertainty about the abbreviation list would be eliminated and professionals would use consistent terminology. It would enable a simple, standard way to link the documents together since they would always be searching for the same list of terms.

Todd and his team are also working to educate the larger Populous community. They are focusing on relaying the successes and lessons learned so processes can be streamlined and all users at Populous can benefit from their developments.

"Specifically, we can do a lot of simple things in terms of drawing and document standards that help facilitate batch linking to avoid the need for post-print magic to make the batch link work," noted Todd.

He also said that quite a few other projects at Populous have also begun collaborating using Bluebeam. One example is the work at Colorado State University with the general contractor Mortenson, who is also a Bluebeam user.

Overall, Todd and his team see Bluebeam as a modern tool that simplifies the everyday workings of a project. While Populous is still exploring what else the software is currently capable of, they're eager to see what other functions and features Bluebeam develops in the future.

Summary

The final batch functionality covered in this section is Batch Summary, the ability to summarize redlines or markups from multiple files in one single process.

Similar to Batch Link, the user can access the Summary function by going to File > Batch > Summary. The window shown in Figure 9-26 will allow the user to select the files of interest. Because the process is fairly intuitive and also very similar to the Batch Link process covered in the previous paragraphs, adding files will not be covered in detail here.

Figure 9-26: Batch Summary Add Files Window

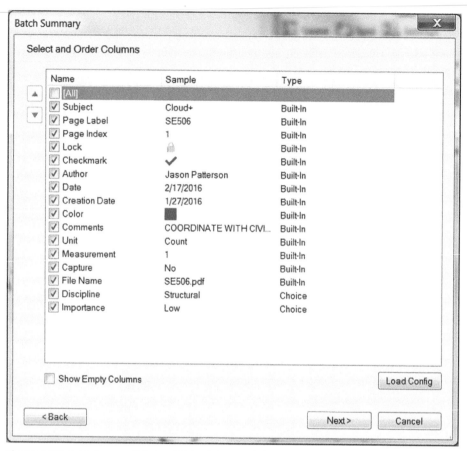

Figure 9-27: Batch Summary Select and Order Columns Window

After adding the files, the user clicks Next and is taken to the Select and Order Columns window, as shown in Figure 9-27. This window displays a list of every markup data column currently in use within any of the files selected. The data in these columns is what is being summarized across all the selected files. The window gives the user the option to exclude and reorder any of the columns in order to meet the users' needs. If the user is only interested in the comments and the author of the comments, he or she can select two columns and click Next to move on.

Before doing so, note the Load Config button in the bottom right. This convenient feature allows the user to recall a previously used Batch Summary setup so he or she doesn't have to re-create it. The Save Config option will be noted later.

The next window is actually where the power of the function is contained. Here the user can filter and sort the summary based on the data within the selected columns. As shown in Figure 9-28, each of the columns is listed in the left column of the table. In the right column, the dropdown menus contain a list of all entries in the given column and allow the user to select the values to include or exclude from the summary.

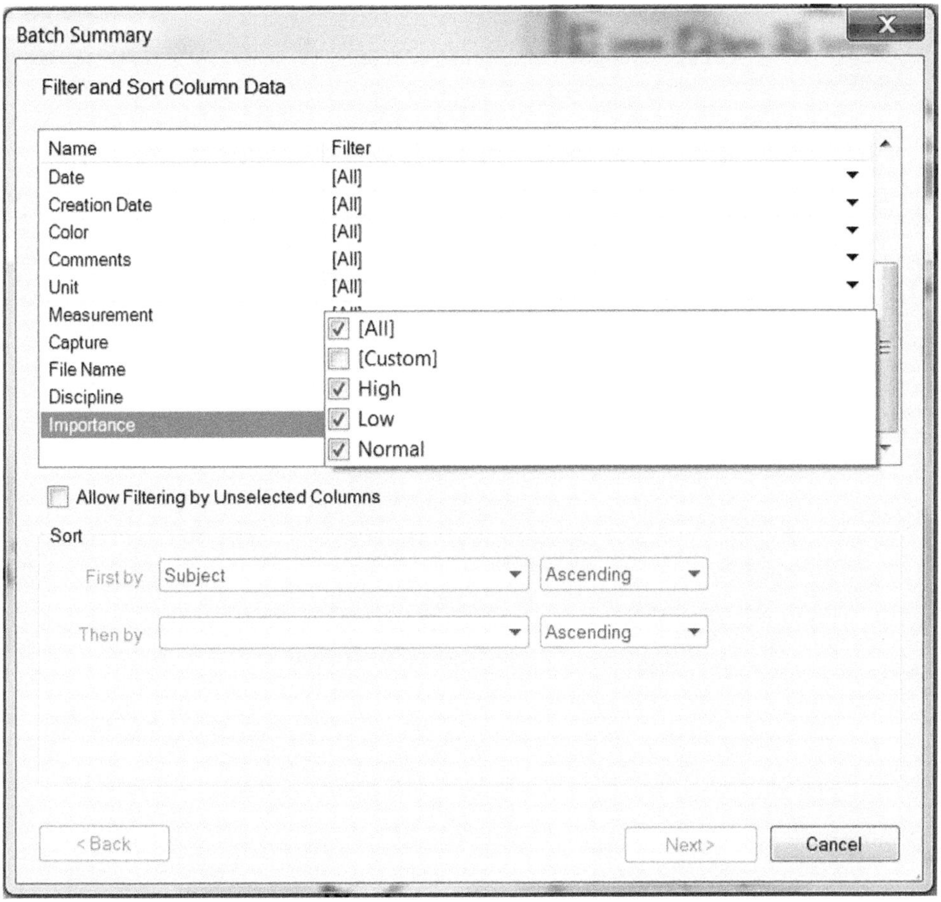

Figure 9-28: Filter and Sort Column Data Window

For example, if the user only wishes to see markups identified as "High" importance, he or she should click the dropdown associated with Importance and uncheck everything but High. When the summary runs, it will only include "High" importance comments. Any comment not meeting the criteria specified in each of the column dropdown boxes will be excluded from the results.

In the bottom half of the same window, the user can specify an unlimited number of sorting options. Sorting simply orders the comments in the summary. A user may want to see comments sorted by engineering discipline so all the mechanical, civil, and similar comments are clustered together.

Upon clicking Next, the user will be taken to the final Batch Summary window. This is the fun window where the user gets the opportunity to customize much about the appearance of the summary.

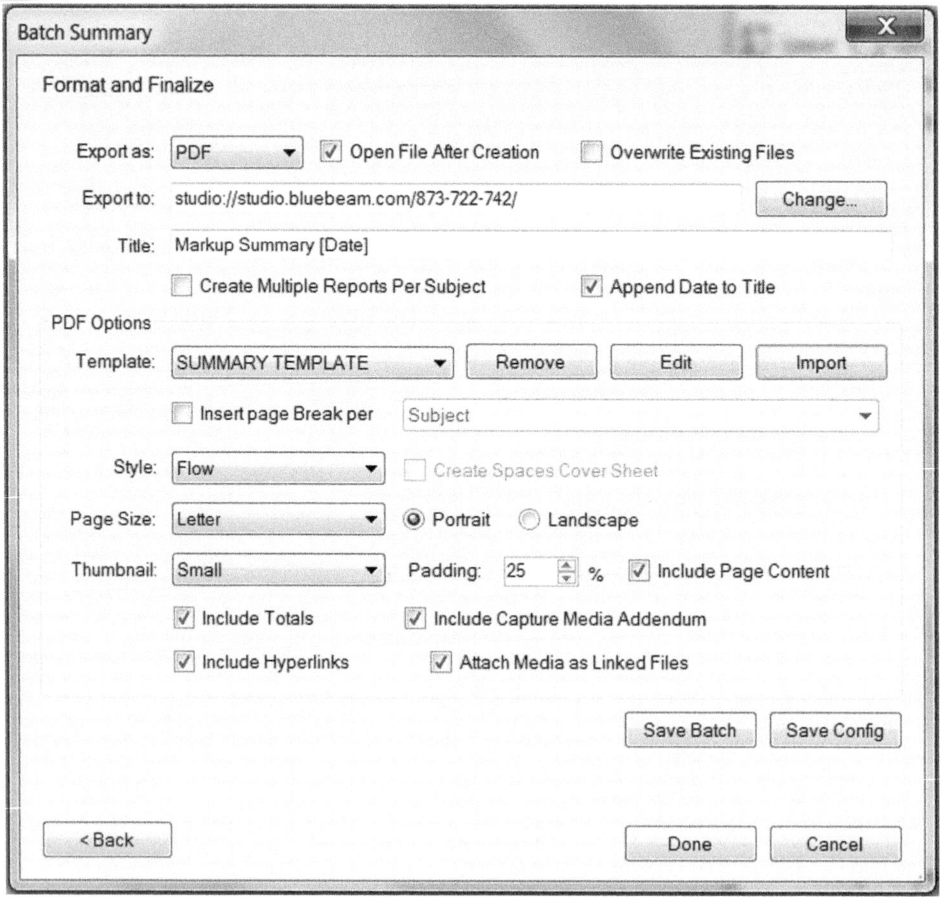

Figure 9-29: Format and Finalize Window

As shown in Figure 9-29, at the top of the window the user may select some basic options such as the export file type, export location, and Title. Two important options are Create Multiple Reports Per and the ability to append a date to the title, which comes in very handy for repeat summaries where the user would typically have to rename the file in order to prevent overwriting the old one.

The first option easily allows the user to break the report by top-level sort category. In the example, that category is Subject, but it could be any of the columns including author, discipline, or importance. This becomes very nice for project managers because they can create a separate report for each discipline without having to run multiple batches.

Below the break, the user has the ability to select a template, the style of the report, the page size, and finally some options concerning the images that will be included in the report. Most of these settings will have to be explored by the user, but the Template option will be touched upon here.

Template allows the user to import an existing PDF document to use as the backdrop of the summary, somewhat like a letterhead. It's an incredibly valuable capability because now users can make these reports look like they are from their company with very little additional work. To include a template, simply click Import and navigate to the desired background file. Once selected, the window shown in Figure 9-30 will appear.

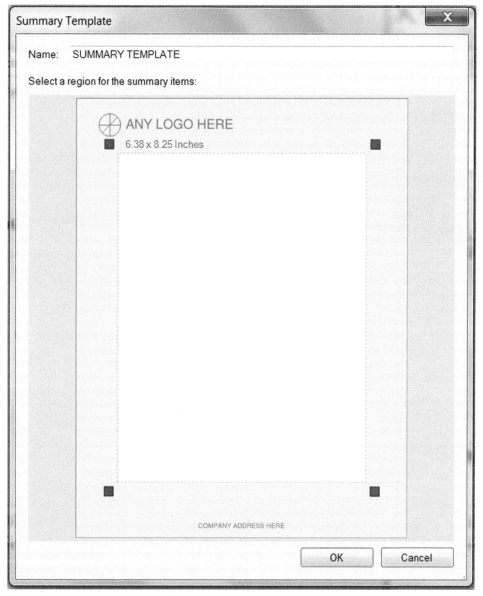

Figure 9-30: Summary Template Window

The user will immediately see his or her template in the background with a white box over the top of the page. The blue handles on each corner of the box may be used to resize the area in which the summary will be printed. Once defined, Revu will adjust the pagination of the summary to fit within the white space nicely. Clicking OK will save the settings.

After adjusting any parameters in the Format and Finalize window, the user may Save the batch for future reuse. Once complete, clicking Done will create the summary file and open it in the appropriate software. A sample report is shown in Figure 9-31.

Markup Summary 4-10-2017

Callout (1)
Measurement: 1 Count

Subject: Callout
Page Label: C-203
Page Index: 1
Lock: Unlocked
Checkmark: Unchecked
Author: Jason Hascall
Date: 2/17/2016 6:41:15 AM
Creation Date: 1/30/2016 7:41:59 AM
Color: ■
Unit: Count
Measurement: 1 Count
Capture: No
File Name: C-203.pdf
Discipline: Geotechnical
Importance: High

CON
SET
PRO

Figure 9-31: Sample Batch Summary Report

Document Security

The final section of this chapter concerns document security. No doubt there are times when a document needs to be protected, whether from being opened, being edited, or being shared. Bluebeam, and the PDF format in general, have two main strategies for protecting a file or the content within a file, Document Security and Digital Signatures.

Digital signatures has been covered thoroughly in Chapter 6 and Batch Signatures will be covered in Chapter 10, therefore this section will focus largely on general document security.

Settings on an existing document can be found under the Document > Security menu, as shown in Figure 9-32. As shown, various actions can be restricted or allowed by changing the settings in the window, depending on the desired state of restriction.

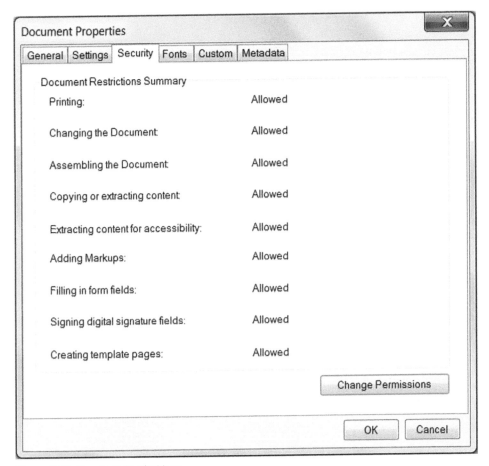

Figure 9-32: Document Security Menu

To do so, simply click Change Permissions from the Document Properties window to display the window shown in Figure 9-33, which is somewhat similar to the previous window. A few things to note:

- The ability to Save Settings. A user who changes security settings frequently can save the settings to recall on future documents.
- The ability to apply two distinct passwords. The first password is for simply opening the document to view it. The second password is for editing and copying.
- The Set Permissions segment is blocked until a password is provided for limiting the printing and editing. If no password is provided, the permissions cannot be changed.

Any changes to the security settings will become effective on the file's next save. From that point on, a password will be required to make any actions that are restricted.

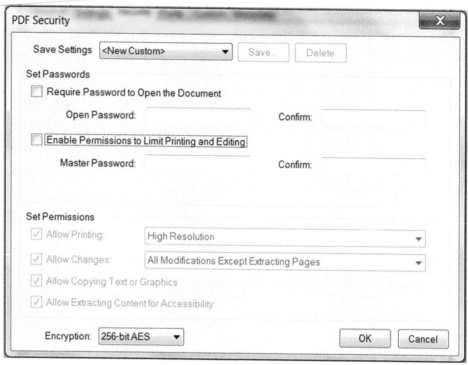

Figure 9-33: PDF Security Options Window

Conclusion

Document Security, along with all of the other features highlighted in this chapter, are instances where Revu is able to assist the user with basic functionality that saves time, increases quality, and simply draws out the benefits of going all electronic. Users are encouraged to utilize as many of these features as they believe add value to their processes and reduce their efforts.

Chapter 10
Go Digital, Engineering

Similar to the previous, this chapter will give a strong focus to the benefits of electronic documents and digital workflows. However, where the last chapter focused on simplifying the navigation of such documents, this chapter will focus on tangible benefits, things that are difficult or simply impossible to do on paper.

Topics include:

- Measuring
- Sketching
- Comparing Documents
- Layers
- 3D PDFs

Measuring

With chapter 7 dedicated to estimating, it's nontraditional to include measuring here; however, measuring is a feature that's much more universal than most of the estimating tools. The authors see it used frequently among engineers, designers, architects, and many other professionals. It's also a capability that is simpler in the digital world than the paper world.

In a typical paper-based workflow, one would utilize a scale to measure a distance on a drawing. The measurement is subject to accuracy issues dependent on the size of the drawings, the accuracy of the scale, and the skill of the measurer. In Revu, measuring a drawing is greatly simplified.

Setting the Scale

There are two ways to set the scale in a PDF document, Define it or Calibrate it to an existing measurement. For correctly plotted drawings, defining the scale is the easiest. As shown in Figure 10-1, most drawings visually indicate the scale of the drawing. When printed on the correct size of paper, the scale defines the measuring parameters.

To define the scale in Revu, simply open the measurement tab ✎ and enter the scale in the boxes, as shown in Figure 10-2.

In this particular example, Figure 10-1 indicates that the scale is 1/16" = 1'-0", so the scale boxes in Figure 10-2 should indicate 0.0625 in = 1'-0" ft' in" as they do. Note that the units

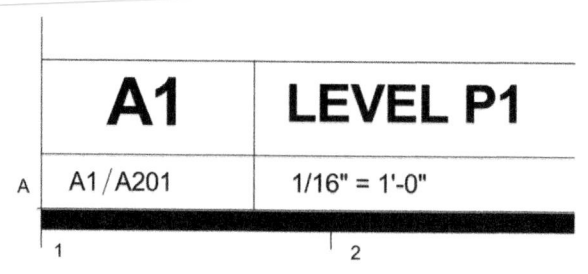

Figure 10-1: Drawing Scale

Figure 10-2: Measurement Scale Setting

in Revu are selectable for each side of the equation. Once the scale is entered, any measurements made on the drawing will utilize the entered scale.

A quick check of an existing measurement in the drawing can validate that the scale is correct, as shown in Figure 10-3.

The second method for establishing a scale is the Calibrate method. To utilize this method, the user simply clicks the Calibrate button shown near the bottom of Figure 10-2 and follows the on-screen instructions. The pop-up box will instruct the user to select two points. Here it's important to choose two points for which the distance between is known. In Figure 10-3, the distance between grid lines 14 and 15 is known to be 36'-0", so the user can select a point on each gridline and enter the distance in the next pop-up window.

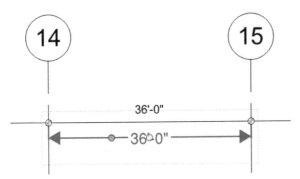

Figure 10-3: Scale Verification

Calibration Contamination

The calibration tool is one of the best tools in Bluebeam and the authors believe it will quickly become one of your favorites. It's great because it doesn't matter if the PDF file was printed on the correct sheet of paper. All that matters is that it was printed in a correct aspect ratio. The tool will take care of the rest based on user input.

However, there are a few caveats and pitfalls to avoid. First, be sure to pick points at a dimension that is known. In Figure 10-3, it's known that grid lines 14 and 15 are 36 feet apart; however, there are points on the two lines that are farther apart than 36 feet. As such, it's important to select two points that are at the same location on each line. Selecting the points at the end of the dimension line is a great practice to fall into.

Second, be wary of viewports. If the screen turns blue or a blue box is indicated, the calibration is likely being set for a viewport, not the entire drawing.

Finally, turn on the snap. Object snap is one of the best features of electronic documents and it's especially handy when measuring because it allows the user to snap to a point of interest such as the end or midpoint of a line. Just be careful to snap to the correct point of interest, not a stray nearby point.

Measurement Tools

Once the scale has been set, the user may utilize any of the tools found in the measurement tab to quickly verify dimensions on the document. Tools include linear measurements, areas, volumes, and even cutouts within areas. Figure 10-4 shows samples of the length, perimeter, and area measurements.

For more information on measuring, readers can visit Bluebeam's website at www .bluebeam.com/us/products/revu/pdf-measurement.asp for a short video demonstration of the tools.

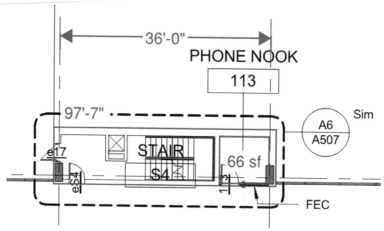

Figure 10-4: Measurement Examples

EXPERT TIP

Function Finesse

As with many things in Bluebeam, finesse has a lot to do with success. In this case, a subtle nuance in the way the area measurement tool is applied changes the way it works. If the user click-hold-drags the mouse, the tool will create a rectangle, orthogonal to the screen. If the user click-move-click-move-clicks the mouse, the tool will create a polygon with control points at the location of the clicks. It's a subtle and very convenient feature of the tool.

Sketching

Even seasoned Revu users don't always believe that sketching can be faster in Bluebeam than by hand, but the authors of this book are confident it's true. Once again, Revu's tricks and features make an electronic method a reality. Sketching is simple, even for those who don't know any of the CAD platforms.

General Sketching

Architects, engineers, and other design professionals often work on similar projects from day to day, week to week, and year to year. They utilize designs in the same or a similar way to what they've created before because they know it works and feel comfortable with it. Often they even have typical or standard details created in CAD or BIM that get used over and over again.

Truth be told, the same trick can be used in Revu. By creating tool sets that capture parts and pieces of typical sketches or details, users can quickly create sketches that could have taken hours. The examples in Figures 10-5 and 10-6 show tool sets containing steel shape cross sections, bolts, and connections and masonry blocks, rebar, and cut lines. Both come in handy by allowing the user to piece together tools into one larger sketch.

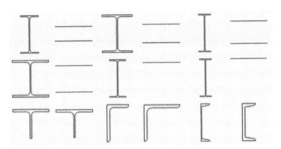

Figure 10-5: Steel Sections Tool Set

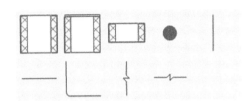

Figure 10-6: Masonry Components Tool Set

A user who begins to sketch frequently in Bluebeam will quickly notice what he or she repeats and can begin to create tool sets appropriately. As discussed previously, these tool sets may then be shared with an entire group to save time and gain efficiencies.

Combine custom tool sets with the standard markup suite and alignment tools discussed in chapter 3 and Revu becomes a well-rounded sketching package. But the fun doesn't stop there: sketching may also be done to scale.

Sketching to Scale

Found under the Markup > Sketch menu, as shown in Figure 10-7, Bluebeam is built with four tools that allow the user to sketch shapes to the scale of the PDF.

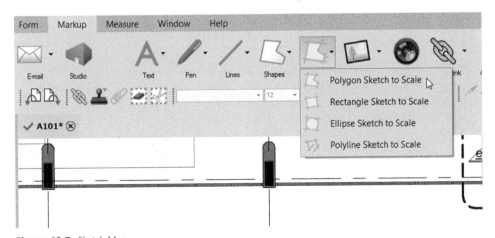

Figure 10-7: Sketch Menu

After selecting any one of the four tools and clicking on the document, the user will notice a measurement box unique to the tools under the Sketch Menu. The user can utilize this box to enter dimensions and rotations of the item being sketched. The rectangle tool includes options to enter the Width, Height, and Rotation, as shown in Figure 10-8.

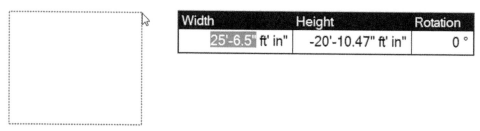

Width	Height	Rotation
25'-6.5" ft' in"	-20'-10.47" ft' in"	0 °

Figure 10-8: Rectangle Sketch to Scale Tool

Unfortunately, as of version 2017, there is no way to edit the dimensions of a sketch to scale item aside from dragging the control points randomly. Nevertheless, the Sketch tools can be very helpful in evaluating competing scenarios where accurate sizing is important.

An obvious caveat with utilizing the Sketch to Scale tools is the need to have the correct scale stored in the PDF. Without the correct scale, the markups will utilize the default scale stored in the file and not be sized appropriately. The scale can be set or modified as indicated in the Measuring section of this chapter.

A short video on these relatively new tools can be found at www.bluebeam.com/us/products/revu/sketch-tools.asp.

CASE STUDY

Meet Ben Naudet and Jesus Rubio, Students and Intern Architects, University of Kansas School of Architecture, Design & Planning

It's easy to say that electronic documents are better than paper documents and point to hundreds of reasons why that's true: safety, security, ease of rework, quality of products, among others. But for most people, the hard-copy habit is hard to break. Society as a whole still wants to print, still wants to hold something, still wants to write. For Ben Naudet and Jesus Rubio, that mindset is disintegrating. Both Ben and Jesus are fourth-year students at the University of Kansas in the School of Architecture, Design & Planning and both have very interesting stories.

For Jesus, Bluebeam started during his first full-time job before going to school at KU. He was part of a small MEP group within a larger engineering firm. The BIM specialist at the firm had the sole license of Revu and spoke the world of it, saying he would never in his life choose another PDF program. Working under the BIM specialist, Jesus too became fond of the tool.

"I was drawn to Bluebeam because it resembled CAD and Revit so much. It felt natural to pan or zoom; and the keyboard shortcuts were just what I wanted. Other PDF software packages simply do not flow as well once you've grown accustomed to the way CAD works," explained Jesus.

But when Jesus left for college, he was also left without a license and his love for Revu was forgotten, at least for a while.

Ben first heard of Bluebeam in the fall of 2015 when he received an assignment in his structures class to trace the structural load paths through his lecture hall. He was given a three-dimensional model of the lecture hall in 3D PDF format and an educational license for Bluebeam. Ben quickly discovered that he could use the 3D PDF to get an explicit view of anything he wanted, which unleashed his creativity.

He said, "I could rotate the model. I could turn layers and elements on or off. It was so helpful because I wasn't bound to a two-dimensional drawing set. I could make literally any view I wanted. It was completely customizable."

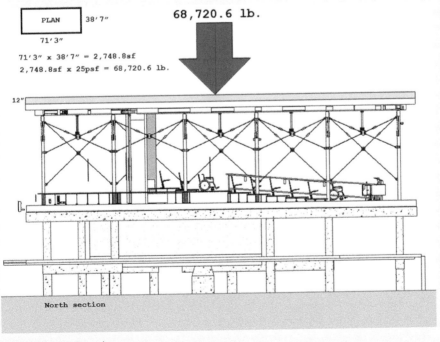

Class Assignment Example

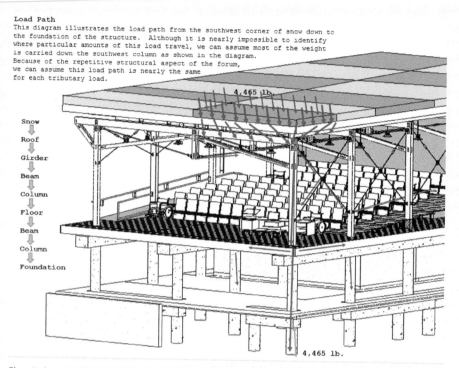

Load Path
This diagram illustrates the load path from the southwest corner of snow down to the foundation of the structure. Although it is nearly impossible to identify where particular amounts of this load travel, we can assume most of the weight is carried down the southwest column as shown in the diagram.
Because of the repetitive structural aspect of the forum, we can assume this load path is nearly the same for each tributary load.

4,465 lb.

Snow
⇩
Roof
⇩
Girder
⇩
Beam
⇩
Column
⇩
Floor
⇩
Beam
⇩
Column
⇩
Foundation

4,465 lb.

Class Assignment Example

As it turns out, Jesus was also in Ben's class, and although he was mostly just excited once again to have access to a licensed copy of Bluebeam, he thought it was cool that the instructor was letting them use this technology. However, he had no idea the capabilities he would soon discover.

As the semester raced on, the instructor introduced the students to Studio, Revu's real-time cloud collaboration platform. He asked all students in the lecture hall to participate in practice problems that were on the screen. Everyone could play at once. Everyone could write on the screen. It was something Ben and Jesus had never seen before and there was a definite cool factor.

Independently, this cool factor caused both Ben and Jesus to explore what exactly was possible with Revu. For Ben, it was the simplicity of the tool that he was fond of. He found it intuitive and was able to take the most basic tools he learned in class and figure out the more complex tools on his own.

"Bluebeam isn't like Adobe Illustrator, which has lots of functionality, but an equally intense learning curve, almost requiring formal training," Ben said. "Bluebeam, by contrast, has a very short learning curve. You can do a lot of basic Illustrator tasks without having to go through the complex Illustrator process."

Recognizing Bluebeam as a combination of the basic functionality found in various Adobe titles such as Illustrator, Acrobat, and InDesign, Ben said, "Bluebeam took a step toward putting lots of other products into one."

Ben's interest grew even more when he was asked to be part of the Bluebeam Revu 2016 Webcast Launch event live from the Nokia Theater in Los Angeles.

"Being involved with the webcast event caused me to go even deeper," Ben recalled. I learned that there's a lot of stuff in Bluebeam and it is clearly an incredible tool."

Beyond gaining Bluebeam skills, Ben had the opportunity to learn a significant amount about the AEC community and meet a number of AEC professionals who use Bluebeam every day in their professional careers, people who would never give up Revu.

The next semester, Jesus and Ben's instructor had his students utilize Studio for the review cycle of the class project. It allowed everyone in the class to comment on everyone else's projects during the same window of time.

"Commenting on projects was good. Studio was an incredible tool because it multi-connects people. It's not a one-to-one. More than one person can be in the exact same file," said Jesus. The use of Studio in the classroom exposed the two to even more of Revu's capabilities.

Today, having taken advantage of the education give-away Bluebeam made possible following its 2016 release, Ben and Jesus both utilize Revu as their default everyday PDF tool. They create PDF files to utilize in their architecture classes. They use the PDF format to share information. They take notes on a touch screen tablet. Jesus even introduced it to the local architecture firm in Lawrence, Kansas, where he interns.

He noted, "My firm is still not using Bluebeam only because they've already purchased other licenses. But editing the PDF content is very appealing to them and I suspect they will go to Bluebeam when they completely transition to Revit."

Jesus showed his boss how the Revu add-in for Revit creates PDFs with details and section cuts that still have live links from their plan views, meaning a simple click will take a user where he or she wants to go. His boss was impressed, and recently agreed to test Revu on a small upcoming project.

"The client didn't want to pay for drawings," said Jesus, "but when we went to the site, we found that the as-built drawings didn't match the existing."

Utilizing a trial license, Jesus was able to show his company that Revu could be used to mark corrections on the existing drawings without recreating the whole set in AutoCAD. The firm went on to utilize those as-built drawings in designing the new additions that were created in AutoCAD. Jesus's solution was a hit. It worked well and made the client and his boss very satisfied.

Ben and Jesus plan to use Bluebeam well into the future.

"I really hope I'm using Bluebeam in my architecture career," said Ben.

Jesus seconded that, saying, "I don't want to go back to the pre-Bluebeam days. Bluebeam is *every day* for me.

He can't imagine how valuable Studio must be on larger projects where collaboration is king and went on to note that at least one architecture firm at the Kansas University career fair was listing Bluebeam as a desired skill.

No doubt, the use of Bluebeam is expanding and firms all around the world will start expecting it to be utilized by their professionals. Lucky for those professionals, Revu is relatively easy to learn, but getting a head start in school is never a bad thing.

Comparing Documents

In a typical design workflow, there comes a point in time where changes to the design have great implications, often causing delays and unexpected costs. After that critical threshold, designers generally cloud their changes to make obvious what is different. This helps the recipient quickly identify what he or she needs to pay attention to.

However, sometimes that luxury isn't provided. The clouds are forgotten, inadvertently omitted, too encompassing, or simply not yet provided because the critical threshold hasn't been passed. During those times, it's hard to feel confident that all the changes have been noticed. For those days, Bluebeam created two document comparison tools, Compare Documents and Overlay Pages.

Both function to illustrate the differences between two or more very similar documents. Each has a specialty and a specific use case. As readers will quickly discover, for every comparison use case, one of the tools will work better than the other, but which one wins is dependent on the specific use.

This chapter will illustrate how to use each of the methods and offer some helpful tips to follow along the way.

EXPERT TIP

Not for Word Processing

Try as one might, the Revu comparison functions are not tailored to word processing documents and will not beat the functionality of software packages like Microsoft Word.

The problem is simple really, when a single word is added to a paragraph in a word processing document, at least the remainder of line has technically "changed" because the words are no longer in the same place. This could compound to the next line or remainder of the paragraph if word wrap caused a word to jump to the next line. Worse yet, it could impact the entire document if the single word resulted in an extra line, which then shifted all remaining text down one line.

Bluebeam, in turn, will identify the entire remainder of the document as having changed, resulting in a very unhelpful analysis.

As such, users should stick with the word processing comparison tools for word processing documents.

Compare Documents

The Compare Documents tool could be nicknamed "Cloud Changes." The tool allows the user to choose two similar documents and then scours through those documents, pixel by pixel, to find the differences. Once the program has identified those differences, it places a cloud over the differences, making the change immediately obvious to the user.

The easiest way to begin the Compare Documents process is to open the two documents of interest. Next select Document > Comparison > Compare Documents and the window shown in Figure 10-9 should appear.

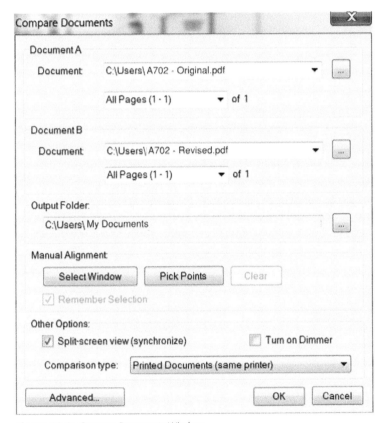

Figure 10-9: Compare Documents Window

Notice the two sections at the top of the window, Document A and Document B. If the two files to be compared are the only two files open, Revu automatically selects the two and assigned one to Document A and the other to Document B. If no files are open at the time the tool is selected, the user will need to navigate out to select the two files to be compared. And finally, if more than two files are open, Revu makes its best guess which files should be compared. In any case, the user should verify the two files selected are correct. Note that the dropdown selection menus allow the user to choose any open file.

Next the user should verify the pages to be compared. Often the documents are the same length and each page of the document should be compared sequentially. In other special cases, page 1 of Document A may need to be compared to page 5 of Document B. Either case can be accomplished by correctly selecting the pages in the pop-up window.

The next segment of the pop-up window is the Output Folder, where the user can select the location to save the newly created Difference file, which will contain all the cloud markups. By default, Revu selects the location of Document A.

The bottom half of the window is where the real power of the tool comes into place. The first box concerns alignment and gives the user the option to manually line up the pages being compared. If the two documents are printed with the same settings on the same size sheet, Revu will easily accommodate the comparison because the information on the two pages will start and end at the same place. If, however, the two documents were printed with different settings or at different scales, the documents will have to be manually aligned.

To manually align the whole page, click the Pick Points box, which will prompt the user to select three matching points on each of the two files.

EXPERT TIP

Pick Smart Points

For the best comparison possible, it's important to follow some general rules when picking points:

1. Select points that can be easily identified on BOTH Document A and Document B. Choose sharp corners or ends of lines that the user knows will be at the same location in both documents. Don't choose vague or ambiguous points such as letters or the center of a logo.

2. Select points that are far apart. Wider is better and selecting points at the extremities of the page can help minimize the error that can be compounded if corresponding selected points are a pixel or two off.

3. Utilize Snap if possible. If the pages to be compared contain vector content, utilize the Snap feature to have the best chance at selecting the same points.

Once the points are selected, Revu will align, scale, and rotate the two pages so those three points align and then complete the comparison function. The other option in the alignment box allows the user to isolate a specific window to compare rather than the whole page.

The final box at the bottom of the pop-up window and the Advanced button have options for the user to explore on his or her own. The choices are fairly intuitive and will make more sense as the use of the tool increases.

Selecting OK will begin the comparison process, which, in the example in Figure 10-9, results in the creation of a new file named "A703 - Original_Diff." This is the Difference file and contains cloud markups at every location where the Document A and Document B are different. Figures 10-10, 10-11, and 10-12 show Document A, Document B, and the Difference document.

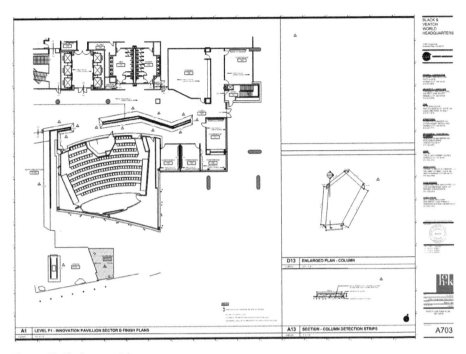

Figure 10-10: Document A

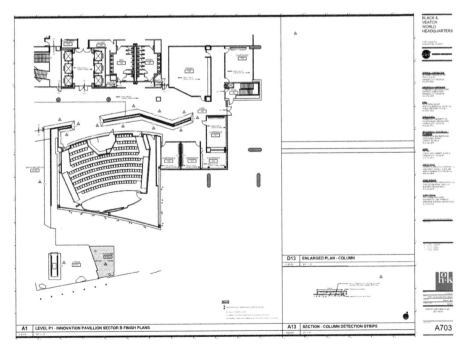

Figure 10-11: Document B

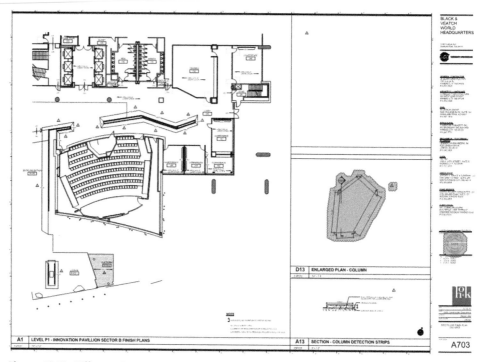

Figure 10-12: Difference Document

Zooming in on the changes as done in Figure 10-13, the reader can see that Revu has clouded the detail D1 and the Architect's seal. Looking back at Documents A and B, the user can easily see that both exist in Document A, but have disappeared in Document B, exactly as Revu has identified.

As an added bonus, Revu's cloud markups now appear in the markups list with the subject "Difference," as shown in Figure 10-14. The beauty of this is that the user can now utilize the list to jump to the differences, or filter out only the differences avoiding other markups in the files. When working with a multi-hundred page file, these capabilities are key.

Overlay Pages

The second comparison tool in Bluebeam is the Overlay Pages tool. For any readers who have been in the industry for a while, this is essentially a digital light table. For readers who don't know what a light table is, a quick Google Image search will show you exactly how one works. Essentially a back-lit table is used to project light through one drawing to another drawing in order to trace the old onto the new or identify the differences between the old and the new.

Revu's version of the light table is digital and converts each of the documents to a monochrome color before overlaying them on top of one another. Once overlaid, the common pixels are turned black and the uncommon pixels maintain their monochrome color such as red or green or blue.

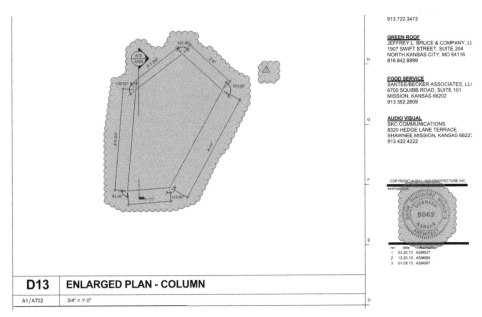

Figure 10-13: Isolated Changes

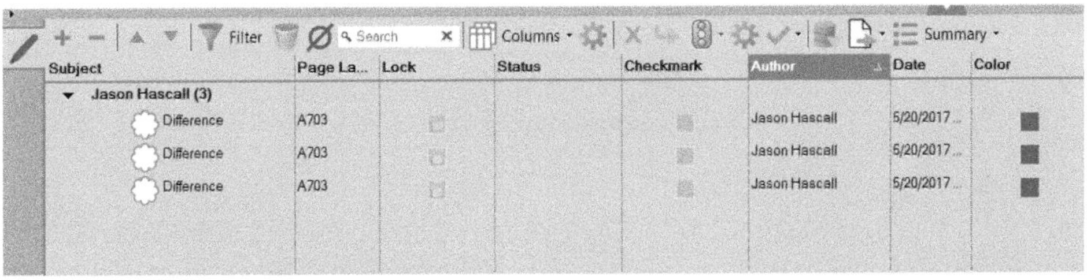

Figure 10-14: Differences in the Markups list

To begin the process, select Document > Comparison > Overlay Pages. The pop-up window shown in Figure 10-15 will appear and list any files that are currently open in the Revu window. In this particular example, the two files utilized in the Compare Documents function were also open here.

The first thing to notice is that each file has been assigned a color. By default, Revu assigns red to the first file and green to the second, but both are customizable by the user. The second thing to notice is that unlike Compare Documents, Overlay Pages is able to accommodate more than two files. In this window, the user has the ability to click Add and navigate to any other files. As files are added, Revu will assign another default color to the file.

The two other buttons in the window are Edit Defaults and Align Points. As expected, Align Points works exactly like it does in the Compare Documents function and the user should expect the same experience. Edit Defaults gives the user the opportunity to redefine the default colors and adjust other settings for the process.

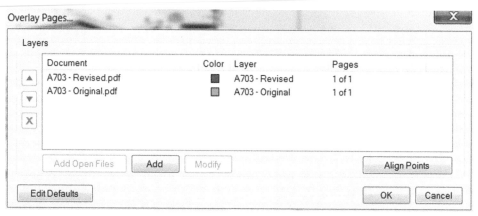

Figure 10-15: Overlay Pages Window

Once the files are selected and the colors are defined, the user clicks OK and Bluebeam creates a new file named "Overlay," stacks the two documents on top of each other, assigning each to a distinct layer, changes all the foreground to the defined colors, and opens the new file.

Figure 10-16 shows the view of the changes identified before. In this figure, the reader can see black borders and text, with a green detail and a green architect's seal, implying that the detail and the seal appear on the A703 - Original file (the file assigned to green), but not on the A703 - Revised file (the file assigned to red).

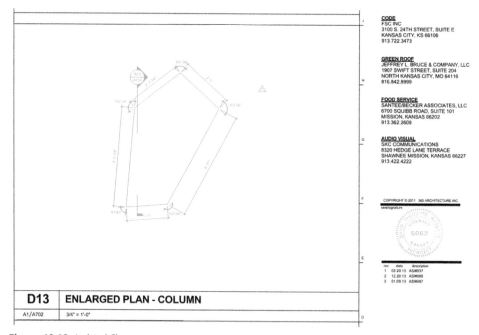

Figure 10-16: Isolated Changes

As an example where both documents have changes indicated by color, consider the two images in Figure 10-17 and Figure 10-18, where the men's and women's restrooms have been reversed. Figure 10-19 displays the overlay file for the two.

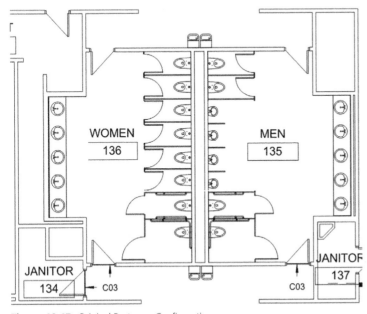

Figure 10-17: Original Restroom Configuration

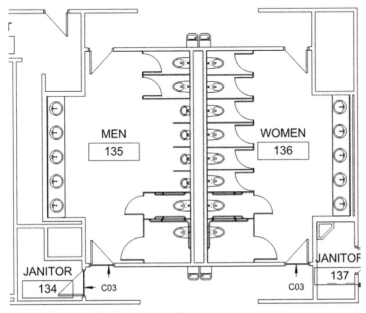

Figure 10-18: Modified Restroom Configuration

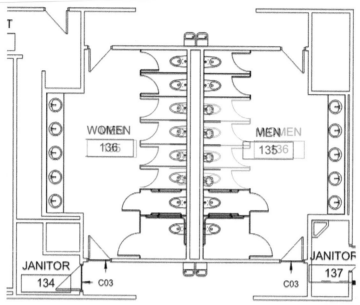

Figure 10-19: Overlay File

In the final graphic, the reader can see red and green linework indicating the individual contents of each distinct file. The reader can also see that outside of the changes, the linework has overlapped exactly and turned black.

Each of the two comparison tools has a place. When the user is comparing Revision A of a 500-page set of drawings to Revision B of a 500-page set of drawings, Compare Documents usually wins the start. Scanning through 500 pages of overlay is time consuming, while a concise list of all the clouded changes is a great benefit, particularly when the changes are minimal.

On the flip side, when comparing the differences between five documents, the Overlay is the clear winner since Compare Documents can only match up two. In other circumstances the best scenario will not be explicitly clear, but likely one will work well and the other will not.

Further, it's not out of the question or realm of possibility to run both on the same two documents, creating the light table effect while also clouding the changes.

Finally, readers interested in seeing more on the two comparison functions can find a short video on Bluebeam's website at: www.bluebeam.com/us/products/revu/compare-documents.asp.

Layers

Layers are a feature that some users swear by and others hardly touch. No doubt, it depends on the use case, but layers have some pretty interesting functionality. This section will cover some of the basics of layers and also touch on a few lesser known capabilities.

The Basics

At the basic level, layers are exactly what they sound like, individual levels that are stacked on top of one another to make a combined whole. Every PDF has at least two layers, neither of which is explicitly defined or identified as a layer. These are the content layer and the annotation or markup layer. The content layer is what can be thought of as the background, the native PDF that was created from another software tool such as Autodesk AutoCAD or Autodesk Revit. The annotation layer is that layer, or level, above the content layer where the user makes his or her markups.

On top of these default layers, Bluebeam allows the user to add any number of additional layers; although many users do not realize it, each of those layers can have a content level and an annotation level.

At the surface, Layers allow the user to toggle content or annotations with the click of a button, either isolating or hiding part of the document. For example, a user may wish to send a punch list to an electrician. With layers, the user can toggle off the mechanical, plumbing, and framing punch items, leaving only the electrical items for the electrician.

Layers could also be used to isolate markups from various users or disciplines. For instance, during a quality control check, users could work on individual layers so all structural engineering comments were placed on a single layer and all architectural comments were placed on a different layer. Why would a user want to do that? Perhaps to send a filtered markup set to a client, partner, or customer who doesn't have the benefit of Bluebeam and its powerful markups list.

Figure 10-20: Layers Tab

To begin working with layers, find the Layers tab, as shown in Figure 10-20.

From the Layers tab, there are two ways for the user to generate a new layer, the Add New Layer button on the far left, identified by the yellow starburst, and the Add Layer from Page button second from the left, identified by the green "+" symbol.

Clicking the first button will prompt the user for a layer name with the box shown in Figure 10-21. After the user enters a name, the new layer will appear in the list. Clicking the second button will prompt the user with the box shown in Figure 10-22. Because creating a layer from existing page content is more involved than simply adding another layer, the box has many more options.

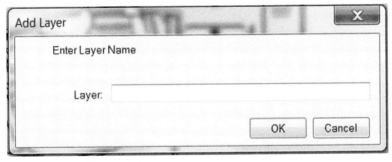

Figure 10-21: Add Layer Window

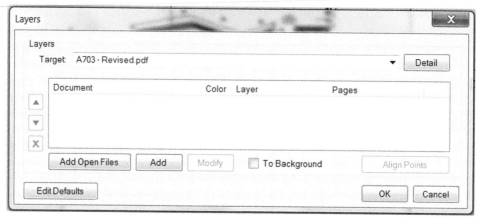

Figure 10-22: Add Layer from Page Content Window

Adding a new layer from page content allows the user to pull content from another page and place it on its own layer in the current document. This can be done as a whole sheet or even a specific page region. Returning to the example used for the compare documents feature, if A703 - Revised was used as the base document, a layer could be used to add detail D3 and allow it to be toggled on or off.

From the Add Layer from Page Content Window shown in Figure 10-22, the user needs to select the file from which the content will be taken—in this example, file A703 - Original. Selecting the file from the list and clicking "Modify," as shown in Figure 10-23, will open a second pop-up window, shown in Figure 10-24, with a host of other options allowing the user to adjust everything from the color and opacity to the scale and rotation. In this example, only the Color and the Page Region selection tool will be used, selecting the region surrounding the detail and assigning it Blue.

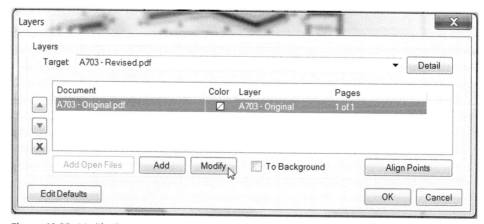

Figure 10-23: Modify File Selection

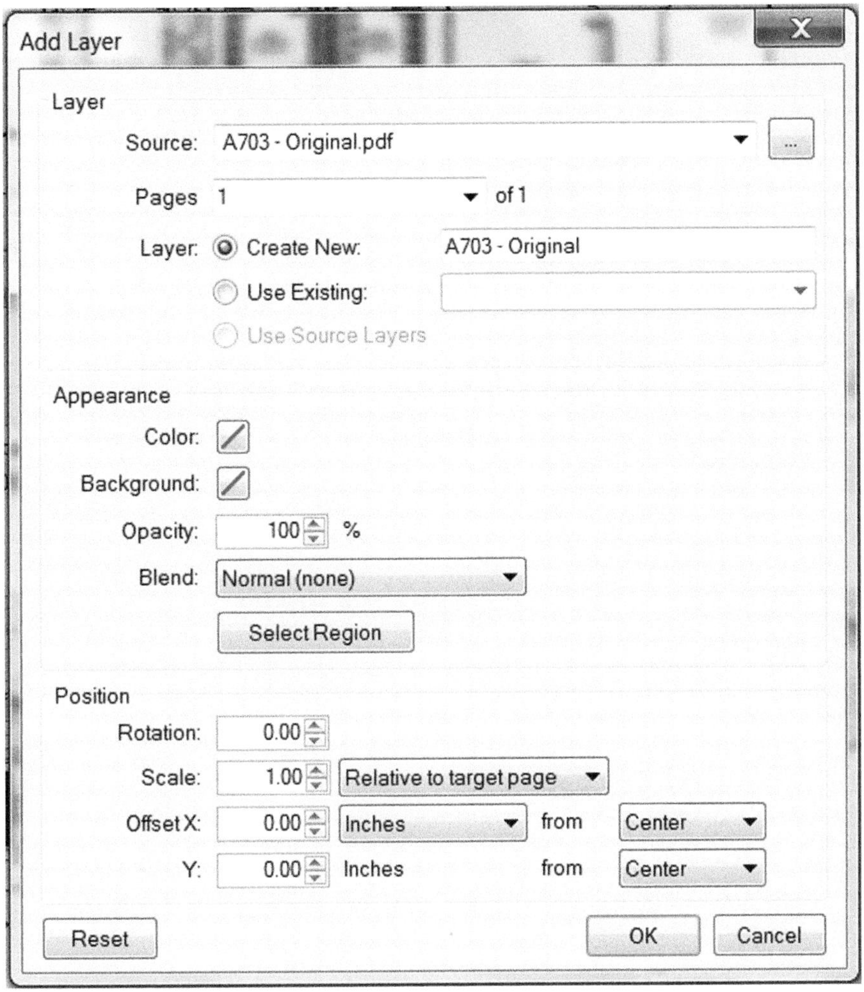

Figure 10-24: Add Layer Modification Window

Figure 10-25 shows the results with the new A703 - Original layer toggled on. To toggle it off, simply click the eye icon at the left end of the layer name. Because the Add Layer from Page button was selected, detail D3 is on a PDF "content," not an annotation layer. As such, it cannot be selected like a markup and it can be manipulated with the content tools such as Cut Content or Erase Content. This is an important discrepancy for the readers.

Layer Assignments

Perhaps one of the best features of layers is that the user may assign tools to a given layer so that those tools will always appear on that given layer.

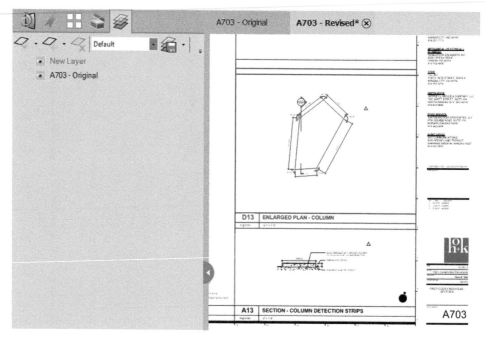

Figure 10-25: Add Layer from Page Results

For example, in structural engineering, steel, concrete, and masonry are typically managed by different subcontractors. If specific punch list or markup or RFI tools are only applicable to one of those subcontractors, the tool may be assigned to a layer representing that trade. To do so, simply right-click on any tool in any tool set and select Layer. A pop-up window will prompt the user to enter the name of the layer. Each time the user applies the tool in a document, the tool will be applied on that layer. If the layer doesn't exist in the particular document, it will automatically be created.

The user may also add any individual markup to any existing or new layer simply by right-clicking on that markup, selecting layer, and choosing the appropriate option, as shown in Figure 10-26.

Layer Behaviors

In addition to the ability to toggle on or off, layers have some additional special features that can be helpful at times. The first is the ability to flatten a layer while still maintaining the layer. Recall that flattening moves information from the annotation layer to the content layer, so any markups existing in a document are pressed into the content layer and no longer appear live in the markups list.

A cloud markup exists in the A703 - Revised document shown in Figure 10-27. Because the cloud also appears as a line item in the markups list, the reader knows it's on the annotation layer and more specifically the "New Layer" annotation layer because the Layer column in the markups list indicates "New Layer."

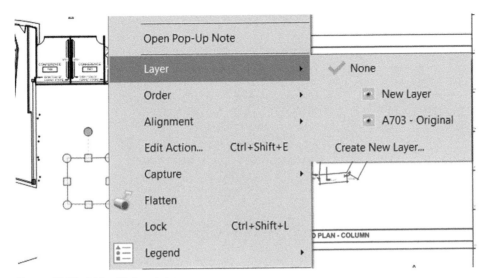

Figure 10-26: Adding Markups to Layers

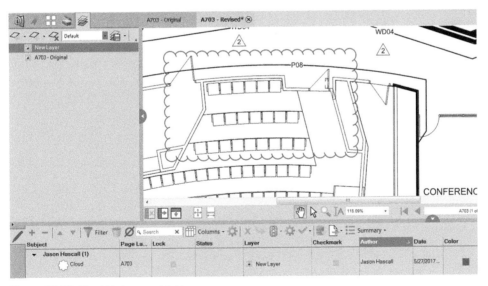

Figure 10-27: Cloud Markup on a Layer

After flattening the document with the Assign Layer option unchecked, the cloud markup has been pressed into the content layer of New Layer and no longer appears in the markups list, as indicated in Figure 10-28. Because Revu still recognizes the layering scheme, however, the cloud can still be toggled on or off with the layer.

This particular feature can be very handy for working with multiple disciplines because it allows quick toggling without risk associated with live markups. It fully enables the subcontractor workflow mentioned previously.

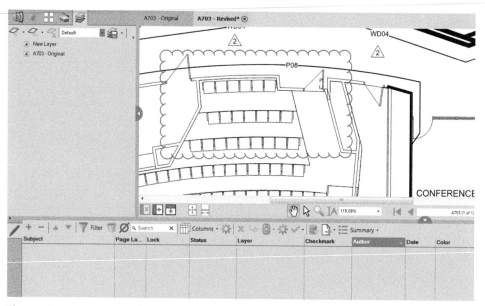

Figure 10-28: Layer Content after Flattening

Other features and behaviors of layers can be tapped by right-clicking on a layer in the Layers Tab and selecting Properties at the bottom of the menu. This will bring up the window shown in Figure 10-29.

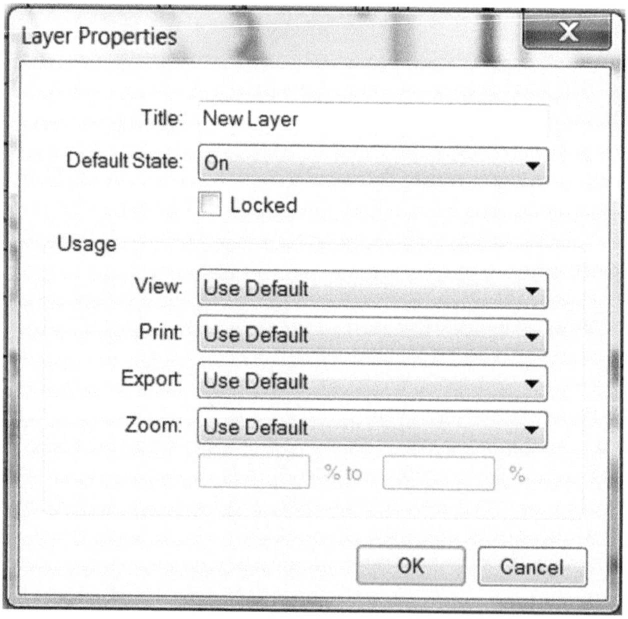

Figure 10-29: Layer Properties Window

Here the user can change a number of behaviors such as printing, exporting, and zooming. For example, the zooming feature can be utilized to toggle layers on or off based on the zoom level in the document. To do so, simply select On Within Range from the Zoom dropdown menu and define the two limit state percentages of the zoom level.

If the limits are set to 75% and 400%, if a user opens a document at a zoom level of 30%, the contents of the specific layer will not be shown. As the user zooms in, the contents of the layer will appear at 75% and disappear again at 400%.

EXPERT TIP

No Print Layers

The reader may remember Larry Naab as a pioneer of using Bluebeam at Black & Veatch from the Case Study in chapter 1. Larry is also a big fan of NO PRINT layers. He cleverly uses these layers as templates for hard-copy print items like CD labels, envelopes, and cover pages.

Larry has long thought that creating such labels and templated items was clunky at best in Word or other software packages. Graphics didn't stay where he wanted them to stay, text wrapped in funny ways, and it simply didn't work. Frustrated, Larry created his own templates in Revu.

He placed circles and rectangles on a NO PRINT layer where the labels were on the sheet. He scattered commonly used logos around the perimeter, in areas that would be tossed after the label was peeled. He even added a locked markup that could be used as a snap to copy the top label to the bottom, guaranteeing that the two labels were identical.

The combination created a quick way for Larry to create CD labels without the hassle he previously experienced.

EXPERT TIP

Zooming Toggle

On the surface, the zooming layer toggle doesn't seem very valuable, but imagine for a second a future where users aren't bound by paper sizes capable of being printed. Drawings wouldn't be broken up with a key plan into multiple piecemeal sheets. Instead, industry goers may utilize one single electronic PDF. When viewing on iPads, tablets, and PCs, there's no real need for individual sheets; the single sheet can show everything.

That's where the zoom toggle shows its value. Instead of showing all the details of that sheet at all zoom levels, the user could utilize the layers of Revit or AutoCAD to assign components and details to turn on or off as the user zooms in. At a 30% zoom level, the user may see only grid lines, exterior walls, and overall details (desks and chairs are probably not relevant at this zoom level).

When the user zooms into 100%, the furniture, mechanical components, and structural members may appear.

To learn more about layers and see them in action, the reader can check out two videos published by Bluebeam, Inc. The first concerns the basics of layers and can be found at www.youtube.com/watch?v=DGoxOAjHHkE and the second concerns improvements made since layers were first introduced and may be found at www.youtube.com/watch?v= jUuZfTwuFEo.

Batch Sign & Seal

Because of its direct application to engineering professionals, the Batch Sign & Seal feature has been included in this chapter instead of being covered alongside general digital signatures in chapter 6 or the other batch tools in chapter 10.

Introduced in Revu 2017, Batch Sign & Seal is one of the newest tools available in Bluebeam. The authors can only imagine it was also one of the most requested tools as digital signatures can be complicated and are the topic of whole books by themselves.

Batch Sign & Seal attempts to simplify the workflow around signing engineering documents and it does a very good job.

Background

In the days of hard-copy submittals, engineers, architects, and other responsible professionals were required to physically stamp, emboss, or seal and sign each drawing they were responsible for. Stamps were almost always rubber stamps that were dipped in ink and pressed on the hard copies to mark the seal on the drawings. That seal would then be "wet" signed by the responsible professional in ink.

Today, many submittals are electronic, rendering the old process nearly obsolete. In fact, moving to electronic submittals causes significant changes to workflows for all the following reasons:

- Although a "wet" signed drawing could be scanned to become an electronic submittal, the resulting signature would no longer be legally binding and the file would be grainy and large. Further, that signature could easily be copied and replicated by anyone with a Windows PC.

- The days of "guarding and protecting" a professional seal are gone. Any somewhat tech-savvy individual with a computer and the internet could, maliciously or not, create any professional seal in the country in less than 10 minutes.

- Simple electronic seals and signatures created by scanning a signed white sheet of paper or signing an iPad or tablet with a stylus cannot be traced back to the signee, are also easily copied, and are not legally binding.

  Thus the need for a digital signature as discussed in chapter 6. Of course, that sounds fantastic, but there are some technical caveats that lead to problems in traditional workflows:

- Documents that have been digitally signed cannot be altered. This makes sense because the signer wants to guarantee that nothing was added or removed from the document after he or she signed it.

- Even though a digital signature applies to an entire multi-page document, by many state laws, each page of a drawing set requires a seal and signature, not only the cover page. As such, the signee is required to digitally sign each page.

- The act of applying a seal and/or digital signature is a change to the document. As such, any signatures occurring after the first digital signature would invalidate the first. Further, the third would invalidate the second, the fourth would invalidate the third, and so on.

- Finally, a hard-copy printed digital signature is invalid.

As a result of all this, the user/signee is forced to split the file into individual pages and sign each page, subsequently submitting individual files to the client, or setup some very specific provisions for digitally signing multiple times in a single file. In either case, there are significant hurdles and those are what Batch Sign & Seal aims to overcome.

Digital Signatures in the States

As of the writing of this book, nearly all the states in the United States accept digital signatures and electronic submittals as a substitute for hard-copy submittals with "wet" signatures. Each state has its own set of rules for applying those digital signatures.

Some states, such as Florida, have written very intelligent rules surrounding the technology. Others, unfortunately, have written the rules as a parallel to the "wet" signature option, which doesn't lend itself well to the technicalities and nuances of digital signatures.

Revu's Batch Sign & Seal makes great strides at meeting the various requirements of the states, and the developers clearly took the laws into consideration, making the tool as beneficial as possible to the users.

That being said, the authors highly recommend that signing professionals study the regulations of the states to verify compliance with the rules and regulations set forth.

Signing a Batch

To sign a batch set of drawings or documents, go to File > Batch > Sign & Seal, which will launch the window shown in Figure 10-30. Add the desired files or load a saved batch. (The files have already been added in the figure.) After the files have been added, the user has the option to save the batch. Will this set of files be signed again and again and again? If so, it may be a great idea to save the batch so it's easy to repeat the next time.

Clicking next launches the large preview window, as shown in Figure 10-31. The reader will notice the left third of the screen has a number of options pertaining to the seal and signature while the right two-thirds of the screen displays a preview of the first file in the batch.

In this example, the preview happens to be an architectural rendering of *The Forum*, a lecture hall at the University of Kansas School of Architecture, Design & Planning. As a brief aside, this facility is unique because it was designed and built by students in the program. The students completed the project in just one year and did the bulk of the work themselves.

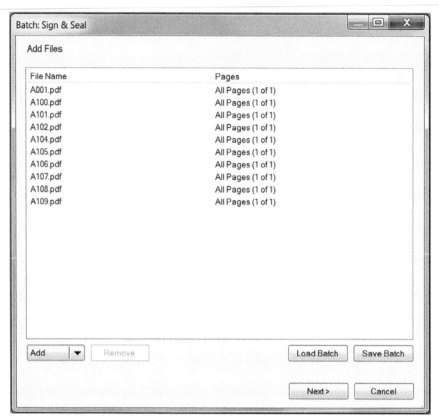

Figure 10-30: Batch: Sign & Seal Add Files Window

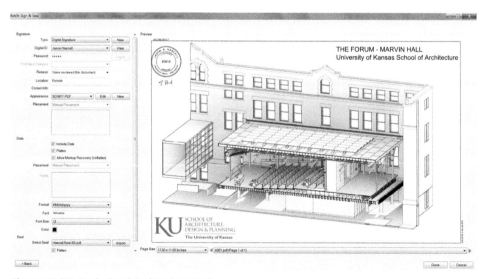

Figure 10-31: Batch: Sign & Seal Preview Window

In this window the user will need to begin by selecting the options on the left. Largely, the Signature section contains the same options that were described previously in the Digital Signature segment of chapter 6. The other two sections, Date and Seal, are specific to this process where it's anticipated that the user is actually sealing drawings or documents. If not, simply toggle off both items, leaving just the Signature.

Advice on the Seal

Through a significant amount of experimentation, the authors have found a very workable technique for the Batch Sign & Seal process and have a few tips to share.

- Utilize a PDF file for the seal. This will allow a transparent background that prevents the seal from blocking any content on the page.

- Do not resize the seal. Many state regulations require the seal to be a specific size. Revu will incorporate the seal at the created size and it should not be resized within the preview window.

- For a "script" appearance that replicates a "wet" signature, utilize a Digital Signature appearance that only has a graphic of the signer's signature. Turn off all the other options. Again, utilize a PDF file format for the signature graphic in order to enable the transparent background.

- Verify vector content. In both the seal and signature PDF files, utilize only vector content; do not use a scanned image of the seal or the signature. The seal can be created directly in Bluebeam or in a native CAD tool and printed in vector format. The signature can be created by signing an iPad, tablet, or other touch screen.

- Choose the flatten options. Flattening the seal and signatures gives a more formal appearance and prevents the recipient from accidentally selecting the seal or date.

In the preview window, the user can manipulate and arrange the seal, signature, and date as desired or as required by state regulations. The user interface allows panning and zooming in and out to more easily facilitate this. Note that the seal/signature/date arrangement will be applied identically to all the sheets in the batch so the user should place them carefully. If specific sheets do not allow placement of the seal and signature in the standard location, the user may toggle through the sheets and adjust them individually. Figure 10-32 shows the final arrangement of the seal, signature, and date.

Before selecting Done, the first-time user may want to save a copy of the files. As with any batch process, that edits the PDF files, Batch Sign & Seal is difficult to undo. After selecting Done, Revu will open each file individually and apply the digital signature along with the seal and date graphics. Figure 10-33 shows the final product. The digital signature is indicated by

Figure 10-32: Seal, Signature, and Date Arrangement

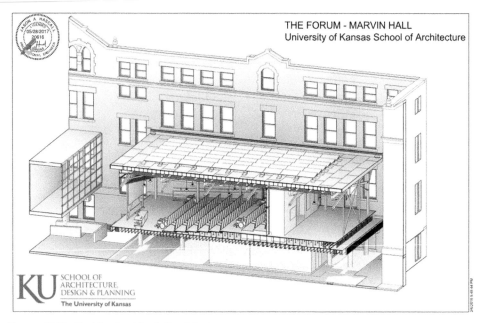

Figure 10-33: Final Results of Batch Sign & Seal

the green checkmark and the date and seal show up just as expected. What used to take hours, opening, signing, and closing multiple documents, now takes a matter of minutes.

Readers wishing to see Batch Sign & Seal in action can find a Bluebeam, Inc. video at www.youtube.com/watch?v=10c8NlBYcOU.

3D PDFs

The final topic of this chapter is 3D PDFs. At the root, these special types of PDF files are a simple and reliable way to share a 3D model with anyone because they can be opened and viewed in any standard PDF reader, which is available for free to everyone.

Revu has taken 3D PDFs a bit further by enabling the markup capabilities and also simplifying the creation of such files by introducing Revit, Autodesk Autodesk Navisworks, and Trimble Sketch-up add-ins. Readers with little exposure to 3D PDFs will almost certainly be surprised and impressed by the functionality available.

Before diving into this section, readers should note that there are many 3D PDF files available for download from the internet. Readers wishing to practice on their own files can find one with a simple Google search, download it, and practice without needing a 3D modeling software to begin with.

User Interface

The general default user interface is shown in Figure 10-34. Readers should note the 3D Hover Bar in the top left and the 3D Model Tree tab on the right. The hover bar contains most of the tools used for manipulating the model, from changing the background color, to slicing, to changing the view type. The model tree contains a full list of all the elements within the model along with associated properties.

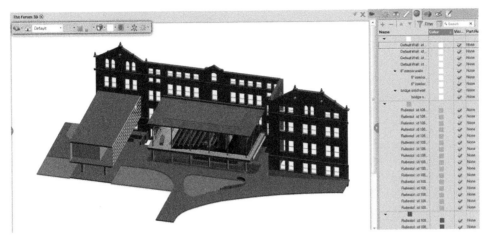

Figure 10-34: 3D PDF User Interface

The first thing readers may want to do is relocate their 3D Model Tree. The aspect ratio of the tab is better suited to be at the bottom of the page rather than the side. Recall that the tab may be easily moved by dragging it from the right panel to the bottom panel. The full-screen width makes all of the 3D tools available without any dragging or scrolling. It also helps display the columns of data in the model elements.

Beginning at the left end of the hover bar, the user will find a selection of cursor functions: rotate, spin, pan, and so on, as shown in Figure 10-35. By default, holding the left cursor rotates the model and holding the scroll wheel button pans the model.

The next three buttons and dropdowns contain the view selections. Creating views will be covered in subsequent paragraphs, but these three buttons let the user select the home or default view, choose an existing view, or scroll through the views as a slideshow.

Figure 10-35: 3D Cursor Functions

Slideshow with a Purpose

The annual Bluebeam eXtreme Conference is a great place for Bluebeam users to pick-up tips, tricks, and new ideas. In fact this tip comes straight from the 3D PDF session taught by Bluebeam.

Readers will see shortly that views can be saved and ordered in the 3D PDF file. This comes in very handy when a professional wants to share a specific vantage point with a client or colleague; he or she can simply save the view and inform the recipient to look for the specific view.

The professional can go one step further, though, and set up a slideshow of views for a client in order to showcase the best, or most concerning, elements of the model. The slideshow can be customized for time and view order, so it's a great way to predictably share information.

Moving to the right, the next four buttons concern details of the view. The first controls whether the view is orthogonal or perspective. The second controls how the model is displayed: transparent, wireframe, solid, and so on. The third sets the lighting projected on the model. And finally, the fourth changes the background color of the window. Readers are encouraged to play around with these and explore how their model can be best displayed.

The next two buttons are where novice users will be surprised, the Cross Section and Transform buttons. The first allows a user to turn on a cross section plane or section box. Both can be used to slice through a model and see what is inside. The cross section plane is shown in Figure 10-36, slicing horizontally through the middle of the lecture hall.

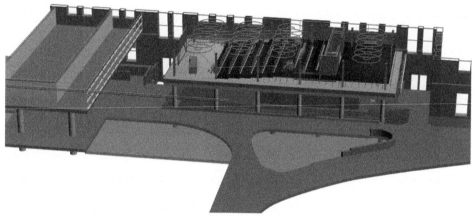

Figure 10-36: Cross Section Plane

To move the plane, simply click on the plane in open space and the directional controller shown in Figure 10-37 will appear. Clicking on any one of the six movement elements will allow the user to move or rotate the plane in that particular direction.

The section box behaves very similarly, but allows users to slice from all sides of a rectangular prism.

The Transform button enables the user to actually move specific elements of the model. In the example shown in Figure 10-38, the roof of the building has been lifted and rotated in two axes. The model elements can always be reset with a simple right-click function, but the transform capability gives the user the ability to showcase things that may be hidden or quickly explore other design options.

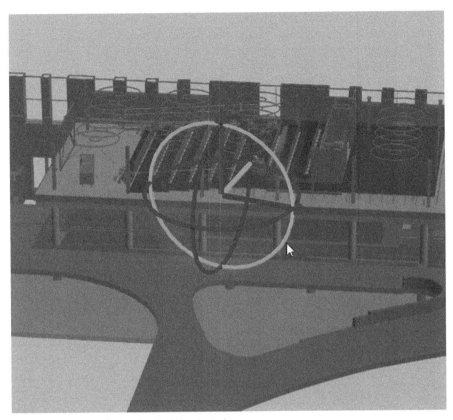

Figure 10-37: Cross Section Manipulation Controls

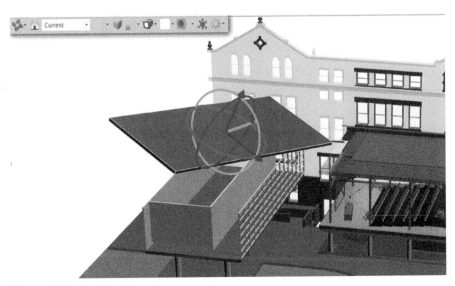

Figure 10-38: Transform Elements

Finally, the last button in the hover bar toggles the 3D markup indicators and the hover bar itself. Readers will see what the 3D markup indicators are in the next section.

Moving onto the 3D model tree, readers will see that the top of the tree contains most of the tools found in the hover bar, but also a few extras. Two additions to note specifically are the Filter, which works exactly like the filter in the markups list, and the View Options, which is where the user can add his or her own views.

The last thing to note about the user interface is that the 3D model tree functions similarly to the markups list. When the user selects an item from the list, the model view will jump to that item and highlight it, thus making it obvious to the user which item was selected. In Figure 10-39, the wheel has been selected and because Transform was toggled on, the user has the capability to move the wheel as desired.

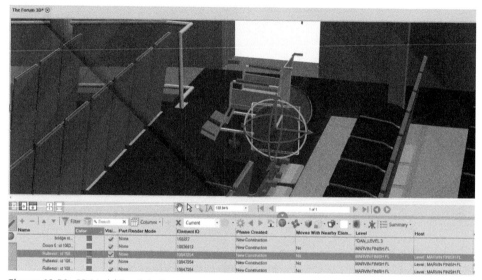

Figure 10-39: 3D Model Tree Selection

3D Markup

With all the powerful capabilities shown so far, it's hard to believe that the coolest feature of Revu's 3D PDF interface may be the markup function.

Annotating any 3D object is difficult because the annotations are always made from a 2D vantage point. Unless tied to a specific element or component, it's difficult to know where in the depth of space a 2D annotation resides. Further, even if tied to a specific element, should the annotation for that element be visible from every vantage point of the model? Likely not.

Bluebeam developers solved this problem by instantaneously creating 2D views when markups are added. In fact, the creation is so seamless that most users don't even realize they are no longer in 3D. In Figure 10-40, a simple callout was added to a 3D view. Note from the view menu in the figure that the current view has automatically shifted to a Markup View, identified by a blue sphere, and the model is no longer able to be manipulated.

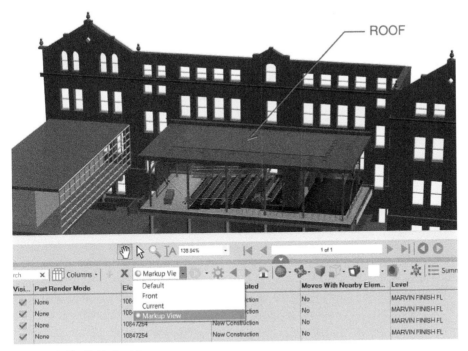

Figure 10-40: 3D Markup View

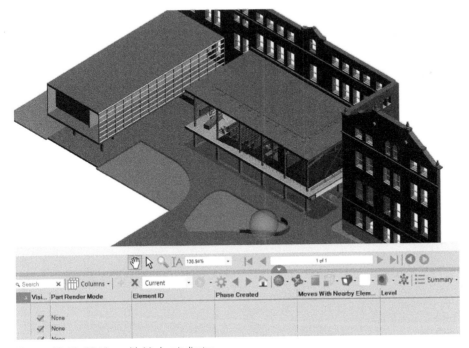

Figure 10-41: 3D View with Markup Indicator

Returning to the 3D view, Current, shown in Figure 10-41, readers can see that the callout markup is no longer visible, but they'll also notice the existence of a blue sphere. That sphere indicates a markup view and clicking on the sphere immediately shifts the corresponding view. The spheres represent individual markup views, not individual markups, and there will always be one sphere for each view. Recall that the spheres can be toggled on and off to aid in model manipulation as they become too numerous.

The final item to note in this basic introduction to 3D PDFs is that the markups being created are still tracked in the markups list. Further, while they maintain their 3D View metadata, indicating which view they are associated with, they are all compiled into one comprehensive list. As such, all the powerful filtering, sorting, replying, and statusing can still be done in a 3D file.

While the industry has been moving to BIM and 3D modeling for quite some time, it seems that 3D PDFs are still in their infancy.

Conclusion

The many features covered in this chapter all commonly enable digital workflows. They, along with the other capabilities outlined in the book, make a truly paperless environment possible for the first time ever. Many software packages have claimed to accomplish this, but, in the authors' opinions, Bluebeam is the first to really make that true.

Chapter 11
Possibilities and Potential

What will Bluebeam developers introduce next? The truth is, the authors have no idea. Only Bluebeam knows what they're working on. However, it is possible to identify some past trends and postulate that those trends will continue. And, of course, one can also dream of what could be possible, consider what features would further change the industry, and imagine what capabilities would simplify AEC workflows. This chapter is dedicated to that, the future. It is complete speculation based on past trends and future needs.

The topics include:

- Recent Bluebeam Trends
  - Document Management
  - Automation
  - Collaboration
- Recent Industry and Technology Trends
  - Data Integration
  - Big Data
  - Single Source of Truth
  - Artificial Intelligence
- Untapped Potential
  - Extrapolation
  - JavaScript
  - Studio API
- Future Possibilities and Potential Impacts

Recent Bluebeam Trends

Beginning with Revu 8, the following list summarizes the most influential features released with each version of Revu. The items belong to three core themes that have continually and recently been woven into Bluebeam's development. Those themes are document management, automation, and collaboration. It's clear that Bluebeam believes each of these is an important strategy for the future of the industry and further clear that the development of the first two themes also supports the third. Collaboration is king and that has never been more evident than the statement made by former CEO and founder of Bluebeam, Richard Lee, in the foreword of this book. Revu is about connecting people through software.

- Revu 8
  - Microsoft Sharepoint & Bentley ProjectWise Integration
  - Autodesk Revit Plug-in
  - Overlay Pages
- Revu 8.5
  - Bluebeam Studio Sessions
- Revu 9
  - 3D PDF Viewing
  - PDF Form Creation
  - OCR
  - Scripting
  - MultiView
- Revu 10
  - Stamps
  - 3D PDF Creation (Autodesk Revit & Navisworks Plug-ins)
  - Bluebeam Vu
  - Bluebeam Studio Projects
- Revu 11
  - AutoMark
  - Sets
- Revu 12
  - Capture
  - Smart Group
  - Batch Link
- Revu 2015
  - Batch Slip Sheet
  - Batch Functions
  - Dynamic Tool Set Scaler
  - Sketch Tools
- Revu 2016
  - Legends
  - Batch Markup Summary
  - Revit Enhancements

- Tags and Drawing Log
- Revu 2017
  - Dynamic Fill
  - Quantity Link
  - Automatic Form Creation
  - Batch Sign & Seal
  - Measurement Enhancements

Document Management

Document management in the industry can be one of the most frustrating and complicated topics in the daily lives of AEC professionals. An engineer graduating from college would never have thought to be spending more time managing drawing revisions than designing, but it's true. Design is only a small piece of what professionals have to do.

Perhaps that's why document management is such a trend for Revu. Bluebeam entered the market with the introduction of Studio Projects in Revu 10. As a lightweight document management tool similar to ProjectWise, Studio Projects gave users the ability to store any number and size of documents in the cloud for free. Revu 10 also brought the introduction of Bluebeam Vu, which enabled Studio participation by professionals without a licensed copy of Revu. Suddenly, anyone with Revu could share files with everyone; it changed the game.

There was no longer the requirement to be on the same network, or utilize slow FTP services. With a simple Studio account a file transfer was easy, and editing the same file from the same source was possible. Further, being hosted in the Amazon Web Service Cloud, security was strong.

Bluebeam continued that trend with the subsequent introductions of Sets, Batch Slip Sheet, Tags, and the Drawing Log, slowly developing Revu to manage, organize, order, and track project drawings and documents automatically. With no more manual effort required, professionals were freed to do what they are best at and also what they love.

But where will this lead? What will Tags enable? What will be next for Sets and Studio Projects? That will be explored in the next segment of this chapter.

Automation

Like it or not, computers are talented. More specifically, they are talented at repetitive, mundane, tedious tasks; the kind that humans are terrible at. At these kinds of tasks, computers are more accurate, faster, and more enduring. Revu developers have starting giving the computer tasks we never thought possible just a few years ago.

Revu 2015, 2016, and 2017 demonstrated more automation than ever before. Revisions of drawings self-identify their matches. Pages name themselves from content within the PDF. Forms auto-populate. And batch capabilities allow users to edit hundreds or even thousands of files in the same time it takes to edit a few.

Once again Revu lifts the weight of the mundane off the shoulders of professionals, freeing them to spend more time doing what they enjoy or focusing on bigger challenges. So what's next? What more can be automated? With the rapid development of artificial intelligence (AI), almost anything is conceivable. Will users simply talk to Revu? Will the interface be more personal? Will the computer step above the mundane and assist with difficult challenges, rapidly iterating multiple scenarios? Will the computer help with creativity?

Collaboration

From the moment the authors began using Revu, the Bluebeam support and development teams were never focused on "one"; they were always focused on "many." Bluebeam needed customer support to make the right decisions in development, but also intuitively saw that customers needed help in connecting with one another, again focusing on the "many" instead of the "one."

The authors believe that sharing and collaboration will soon be the only road to success. For centuries, companies have guarded their secrets, protecting them from their competitors, but today technology is advancing so rapidly that the secrets being protected are outdated before they can be hidden. A company that spends its time and money guarding its tools and processes instead of investing in new tools and processes may quickly find itself irrelevant and starving.

Although nearly every single new feature could arguably be classified as a collaboration development, the largest step toward collaboration was the introduction of Studio in Revu 8.5.

To this day, professionals who are new to Bluebeam are still awestruck by its capabilities. They simply have never seen anything like it before. Competitors claim to have real-time markup capabilities, but in their claim, real-time is every time a user saves the file. We all know that could be minutes, hours, or even days; a far cry from the two-second delay between a professional in India and a professional in the United States. With Studio, real-time is instantaneous.

But tools like Studio aren't just for show; they are allowing people to connect without delay and without a rigid schedule. They are crossing boundaries of time and space. They are throwing away the slow and ponderous waste and encouraging swift and effective execution.

After connecting countries on opposite corners of the world with a mere two-second delay, it was hard to imagine what could be next, but Bluebeam did not disappoint. As if the software developments weren't enough, the team established local communities called Bluebeam User Groups, or BUGs, in major cities across the United States, once again supporting the idea that the "many" is better than the "one."

Recent Industry and Technology Trends

Trends in industry and technology have also undoubtedly impacted the AEC community; and although the authors make no claim of being experts in these trends, they have noticed the following and seen the impact of each directly in their careers.

- Data integration: The sharing and incorporation of a single data set by more than one application. The aggregation of multiple capabilities into a single package.

- Big data: The development of data capture tools, sensors, and analytics acknowledging that data is valuable.

- Single source of truth: The use of a single, shared set of files, avoiding costly delays in distributing current information and expensive rework resulting from multiple versions of the truth.

- Artificial intelligence (AI): The utilization of machine learning techniques to develop mass computing capabilities not previously possible.

One by one, this section will take a brief dive into these topics to see how these broader trends have shaped the industry.

Data Integration

Nowhere is data integration more apparent than in Building Information Modeling (BIM). Just a decade ago, any information about a building was contained in sets of drawings, specifications, calculations, and narratives; separate volumes of information; discrete elements requiring assembly for the full picture. Today BIM is moving toward a future of containing all of that in one single model: structural steel specifications, mechanical equipment part numbers, architectural elements—all found in one place, all stored in an easy-to-navigate container.

Another example is integration of tools with BIM. Structural analysis software packages now communicates directly with the BIM model, assuring that changes made in the construction package are correctly reflected in the structural model. In some cases, the analysis software is even built right in.

The BIM model can also be shared between partners. It can be passed on to contractors to use for construction. It can be shared between designers to evaluate space needs and clash issues. It can be provided to facility owners for building management, maintenance, and operation.

Like never before, the industry is seeing one set of data be reused again and again. Does Bluebeam result in valuable data that should be retained and shared? Can Revu utilize other data that may be valuable in PDF format? A glimpse of this was evident in the Revit integration that transferred preestablished Spaces directly to Revu and also in the introduction of Quantity Link that passes take-off information downstream in the workflow process.

Big Data

Increases in computing power have made it possible to look at mass quantities of data to identify trends that would have gone unnoticed just a few years ago.

Shopping habits are now tracked by Google, Amazon, and others. Cell phone movement can be used to evaluate traffic flow. And software performance feedback can be used to fix bugs and make improvements.

Big data is everywhere, including the AEC community. In the wake of an ever-growing number of available sensors, the data that the AEC community captures is increasing at an exponential rate. In fact, the data stored by the world is more than doubling every two years.

With that comes the analytics, too. Heat sensors in a power plant can now be used to predict a turbine failure well before the plant experiences a catastrophic shut down. Just like

Google, engineering companies can use mass quantities of data over short and long time periods to identify trends, predict outcomes, and build strategies.

Although not on the same scale, Bluebeam is not so different. Comments, tags, measurements, and so on are merely pieces of data that Revu is managing, and further, managing very well. But, what other data can be contained in a PDF? What other data is currently untapped?

Single Source of Truth

Network connectivity and cloud computing have already changed the world. Anyone with a cell phone and a data plan can reach the internet; no cables, no plug-in, no traditional computer required. That connectivity has made a single source of truth possible.

Prior to computer networks, a construction project in the jungles of Indonesia received updated design drawings around once every month. Unfortunately, when they made it to the site, they were already about three weeks out of date. It was a simple fact that it took time to produce copies, compile, and ship drawings to a remote location.

Today, with the advent of cellular and satellite internet communications, today's drawing is the same in Indonesia as it is in the home office. Network connectivity has given the industry the ability to be continuously up to date.

The implications of this are pretty obvious. If a modification to a structural foundation occurs, the foundation may have been built before the drawings indicating the modification ever made it to the project site. At that time, the foundation has to be torn out and reworked or reevaluated and modified, both of which are costly to the company.

Multiple sources of the truth can also cause trouble in the office. When John makes comments on copy A of the design documents and Jill makes comments on copy B of the design documents, there is a risk that John or Jill's comments will not successfully be captured since there are now two sources of truth. Further, working independently, John and Jill don't have the opportunity to see each other's comments, so silly comments about misspellings or typos often get duplicated and subjective comments often conflict. Again, all three scenarios cause delays and additional costs.

Bluebeam Studio and Sets have made great strides in facilitating a single source of truth, but can Revu go further?

Artificial Intelligence

The final topic is a buzzword of today. While long thought of as science fiction, artificial intelligence (AI) is quickly becoming reality. Though not yet as sophisticated as Star Wars' R2-D2 and C-3PO, Clarke's HAL 9000, or Ender's Jane, AI is now capable of much more than tedious tasks.

Machine learning seems to be the outset of AI and has been the engine behind the facial recognition capabilities of Facebook, the Music Genome Project utilized for Pandora, and of course the victory over world champion chess player Garry Kasparov. In fact today, a computer can even create a new Rembrandt-style painting simply by studying existing paintings.

Machine learning is different than human learning because when one computer "learns" a skill, or solves an algorithm, every computer is capable of that same level of skill. Contrarily, each human must learn each skill from the ground up. Further, the first computer developing the algorithm can learn faster than a human because the computer can process faster and there are no fatigue limitations. Computers can run 24 hours per day, 7 days per week, thus allowing them to rapidly iterate on possible situations and outcomes until the perfect algorithm is defined.

AI has nearly infinite potential to change the AEC industry and Revu. Will we be speaking instructions to Bluebeam rather than using a keyboard and mouse? Will Revu be identifying issues for us? Will AI create new features in Revu based on user trends?

Untapped Potential

Beyond considering what Bluebeam might develop, it's important to also consider what users may do with the existing capabilities, how they might script, code, or extrapolate what they need. Readers will see a great example of this in this chapter's case study.

Extrapolation

Each year at the annual Bluebeam eXtreme conference, the Bluebeam team puts together a competition for its users. The idea of the competition is to prove your skills, but the challenge of the competition doesn't have anything to do with using Revu in a traditional way. When the team develops this competition, they have to be pretty creative or they'll give an unfair advantage to a team skilled in one particular area. So, they force teams to solve problems they wouldn't normally solve with Revu using the tools that already exist. There's no better example of doing something nontraditional with Revu.

In one example, Sets were used to stack pages of a specification. When completed, the set contained all the specifications on a project, but the stacked revisions weren't revisions at all, they were individual pages of a given specification. A unique use of Sets could also be the tracking of construction progress photographs. A field professional could take one photo at each of 50 locations every day. The pages in the set could be each of the photo locations, which of course would be hyperlinked to a map or plan view, and the revision stack would be the historical record of the project.

Another example was the use of a PDF as a field weld map. On a past project, weld inspectors were evaluating the damage to a series of dam gates along the Missouri River. With thousands of welds on each gate, it was difficult to track which welds had been inspected, which welds were in good shape, and which welds were in need of repair. Utilizing the existing drawings, designers created a weld map, highlighting each weld on the gate as a markup in PDF. They then modified the Status model from the default Accepted, Rejected, and so on to a custom model, indicating the result of the inspection, Visual, NDT-Pass, NDT-Fail, and the like. The custom model included color change that would differentiate welds that had not been inspected and those that failed inspection. Not only did the statusing keep very good record of the inspection, tracking both time and inspector credentials (user), it provided a sort of "heat map" for the overall status of the inspection and a to-do list for the inspector.

A final example is the composite plan. So many times floor plans are too large to convey any valuable information on a printable size sheet, so designers break them into smaller plans with the addition of a key plan. Yet every construction site in the country has these individual plans taped together on the wall into one big composite plan. In a world where electronic documents are quickly replacing paper documents, why be bound by printable space? Global engineering and construction firm Black & Veatch is starting to create composite plans, or "megasheets," for internal use on some of their projects. The idea is so popular that contractors are now requesting "megasheets" on their jobs.

So the question is, what can users do with what's already out there just by throwing away our typical boundaries and limitations? Can batch link be used to track and connect RFIs, submittals, and other construction documents? Can batch markup summary be used to track project progress? Can layers be used to reduce the quantity of sheets and improve quality of deliverables?

JavaScript

Ten minutes into the annual JavaScript training at the Bluebeam eXtreme conference, all the users in the room suddenly realize the power and potential of JavaScript. It's practically infinite, but it requires development.

One of the most basic applications of JavaScript is the shop drawing stamp. Developers at Bluebeam are kind enough to help Revu users with this one as a means of encouraging them to go further. Readers know the shop drawing stamp as that rubber stamp with paragraphs of legal disclaimers and a handful of check boxes indicating whether the shop drawings are approved, rejected, or somewhere in between. The Java stamp is a super-powered stamp that launches a pop-up window where the user can choose the check box option he or she desires before placing the stamp. Subsequently, the stamp dynamic text functionality captures the username and date. Stamping shop drawings electronically is a cinch.

JavaScript can also be used with forms to populate data from internet sites. For example, a construction manager usually has to complete daily reports. He or she could utilize JavaScript to capture any number of data pieces from the weather and temperature outside on site to the latest image on a webcam.

The eXtreme Conference is a great place to keep up with the latest trends and capabilities in JavaScript. Not only are there specific training sessions on the subject, there are also almost always case study sessions from industry leaders showing their cutting-edge use of the tool.

CASE STUDY

Meet Lane Pemberton, Project Support Specialist, Air Analysis of Atlanta

It's unusual for a company to hire a high school intern, but Air Analysis of Atlanta, Inc. (AAI) took a chance and, as it turned out, the decision was a good one. Lane Pemberton quickly put his innovative spirit to work and began having a positive impact almost immediately.

Lane was first introduced to Bluebeam Revu during his first week with AAI. That week was filled with a solid forty hours of Revu tutorial videos and nothing more. The tutorials, most of which were available for free online, covered a wide variety of subject matter, from basic capabilities to advanced functionality. He learned a lot, but more importantly he gained a respect for Revu as a tool that could do more than he expected.

Lane Pemberton Writing JavaScript

After that first week, his company began handing him simple tasks like organizing files and relabeling documents. Lane did well and soon took on more complicated tasks like scripting in Microsoft Excel. Shortly thereafter, his immediate supervisor, Dane Richards, the Business Operations Manager at AAI, brought over some Excel forms and asked Lane, "Is it possible to make these in Revu?" For Lane, that was the green light he needed to explore what was possible.

Dane was known as the innovator at AAI. He was always pushing the limits with technology and never seemed to run out of new ideas. Seeing Lane's capabilities in Excel, Dane knew he had found the right guy for the job. Lane's scripting was not only valuable, it was complicated yet elegantly simple. He thought it was time to give Lane a chance to show exactly what he was capable of and make his latest idea a reality.

Over the years, AAI had accumulated numerous forms in Excel. They had hit the ceiling with what was possible with scripting and were confident there was no way for them to further simplify or reduce the number of forms they were using. That wasn't acceptable for Dane. He wanted something simpler, something easier to use, something more streamlined. He wanted to start and end entirely within a PDF.

Over the next year and a half, while pursuing an associate's degree in Emerging Technologies, Lane was able to develop 32 scripted PDF forms to do exactly what Dane had asked.

"Our company provides specialized services in the construction industry where our field technicians take measurements associated with airflow, pressures, electrical data, and so on," noted Lane. "The information is gathered from a wide variety of mechanical equipment, so the forms have to be able to document information about all these different types of equipment."

Each of the forms that Lane created applies to a particular category or action of equipment, like air moving, for example. All types of air moving equipment are housed in the same PDF form. Lane has scripted each form so the content of the form dynamically changes based on the equipment type selection made in the header of the form. With each type of equipment selected, the form fields, the corresponding labels, and the data requested from the professionals change corresponding to the selection.

Example JavaScript

Once the correct form is selected, professionals on site fill in the fields with the appropriate measured data and the forms automatically complete any necessary calculations. Beyond just completing the calculations, the form fields also control the format of the data entered, assuring that each of the entries is inside the proper range, measured to the appropriate accuracy, and provided as the correct type. The form will alert the user when the entry does not meet those criteria.

Lane also enabled note functionality, which is where his scripting gets really impressive. Each field blank has a note icon where a user can click to launch a dialogue box where he or she can type supplemental comments, notes, or important information from the site. "A professional may identify an issue with some piece of equipment or how it is operating," he explained. "For example, the measured airflow might be less than what was designated. The professional will provide a comment regarding the issue by clicking the notes icon for the associated field and selecting 'New Note' to enter any necessary information."

More impressive than the ability to add notes is the scripting Lane has done to compile and summarize the notes. In his latest version, when a note gets added, it receives a key number that is displayed directly on the form. At the bottom of the form, all the notes that have been added are listed in the sequence of the key numbers. The key numbers are conveniently ordered by appearance in the form from top left to bottom right. If a note is added between two existing notes, the scripting will automatically renumber all the existing notes in the document based on their location.

AAI also has a note library of common phrases field professionals like to use. They can simply choose a note from the list or type their own custom note. The note library was developed to help build some standardization across the company. Lane added, "Our standard notes allow for consistency across field personnel and account for proper wording and spelling."

New notes are added to the database periodically after a prudent vetting process. Although the vetting process may seem extreme, AAI tries to keep a careful balance between too few phrases to be valuable and too many phrases to be usable.

Finally, once in a while the forms require customizations. Some clients want the forms to look different or provide something specific to that client. For example, a particular client may want an additional column of data that other clients don't need. As such, Lane built in a way to make some customizations like adding the column. He also built in a way to override the automatic calculations if necessary.

But the proof was in the pudding, and Lane's forms hadn't yet hit the field. He had to put them to use and that was harder than creating them. Even with their incredible capabilities, change is hard for people. "We built the forms to resemble the look of the previous paper forms that everyone was used to. We coordinated with some personnel who worked on projects in our area to beta test the new forms and provide feedback. We also obtained feedback from personnel who work in other areas as well as office personnel at review meetings." Lane indicated that there was positive feedback as well as ample amounts of constructive criticism. "Once we incorporated everyone's feedback, we put the forms to use company-wide and even trained some personnel on laptops out in the field."

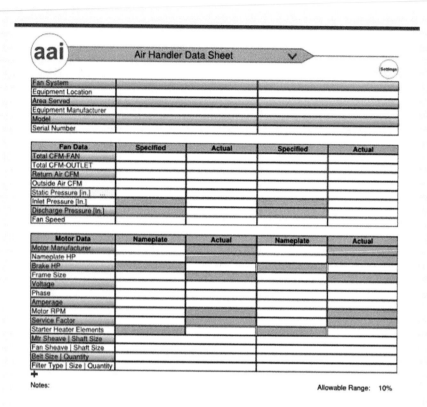

Example AAI Form

After all the hard work and collaboration, Lane was glad to have one of the seasoned professionals in the company tell him the forms were "awesome." The general feedback from the users has been that the more they use the forms, the more they like them.

Looking back at the old way, there are more benefits to the new way than simply ease of use. The paper forms AAI was used to resulted in sloppy handwriting, were easy to lose, and were hard to manage. Field professionals usually had to copy the field notes onto a clean handwritten copy so the notes would be legible for the next professional in the process. That copy was then used to transcribe the notes into the electronic Excel form, which was finally converted to PDF to send to the client. With Lane's forms there is only one step and the form goes right to the client.

"Our clients really appreciate our product. They consider the software and the work we do to be top-notch. Some have seen the benefits that Revu afforded us and requested that we train them how to use it because they see ways in which it can benefit them," Lane said.

With so much success in JavaScript, it was incredible to learn that Lane is a self-taught scripter. He had no Java skills before coming to AAI and learned everything he knows from the internet and trial and error, with a heavy emphasis on the latter. Lane says he has enjoyed the projects he's done at AAI, especially learning to write JavaScript, but more broadly, he looks for any way to be more innovative and takes every opportunity he is offered to do so. He hopes the next step at AAI will be enabling the forms he has created on iPads to make them even easier to fill in the field. Unfortunately, the JavaScript apps are not currently robust enough to do that, but he's confident they will be within the next few update cycles.

Lane also enjoys teaching. In the same spirit as this book, he has no hesitation sharing his knowledge with others. From presenting at the Bluebeam eXtreme conference twice, to working directly with the AAI interns to build their scripting skills, Lane agrees that working together and sharing knowledge is the best way to make progress as a company and a society. Though he knows teaching others might be giving away his unique skill, he's absolutely confident he'll be able to put his innovative spirit to good use in a new and different application. Lane is excited to see what the future holds and looks forward to being a part of it.

Studio API

Creating the Studio API was a big step for Revu developers. For the first time ever, instead of waiting for Bluebeam to develop integrations and tools, developers around the globe could tap into Bluebeam Studio features directly. Not only does this open the capabilities for companies to play on their own, it frees Revu developers from spending so much time building and maintaining individual integrations like the one created for ProjectWise.

Bluebeam gives three great example uses on its website at http://developers.bluebeam.com/articles/example-workflows/.

The first example use of the Studio API is the automatic creation of Studio Sessions from a document management system. Perhaps a company utilizing ProjectWise wants all documents with the tag "QC Ready" to be placed in a Studio Session automatically with no additional action by the user. The Studio API can do that. Another example provided is the backing up of Studio data, assuring the security of a company's information.

The final example is perhaps the most creative, and that is markups list reporting. The Studio API can be used to extract data from the markups lists automatically. That data can then be aggregated into reports and shared with project managers to keep them abreast of their project's progress, a great use of technology to provide information to improve the chances of project success.

Future Possibilities and Potential Impacts

The big question, of course, is where will Bluebeam go next? As stated in the introduction, the authors don't know, but there are a few features they hope to see become part of the package soon. The final section of this final chapter takes a look at three of those developments and why they would be game changers.

Studio + Sets + Status

Any reader can tell that the authors are big proponents of Studio and Sets. Both simplify daily workflows, offer new ways to collaborate, and allow professionals to spend more time on important tasks and less time managing paper. But why not get the best of both worlds?

Bluebeam Sets already function inside of Studio Projects, but the Session functionality isn't quite streamlined. Take Digital Slip Sheeting, for example. Imagine a design team is utilizing Sets with its documents hosted within a Studio Project. Of course, everyone on the team appreciates the ability to mark on the same working document simultaneously, so the project manager uploads each file in the set to a Studio Session.

When it comes time for a revision, the project manager closes the Session, brings the files back into the Studio Project, performs the Digital Slip Sheet operation, and uploads the current and revised sheets back into a new Studio Session.

The problem is, those marks were not created in the Studio Session. Following its default behavior, marks that were made prior to upload are locked from editing by all users. Even if User A created the mark, User A cannot modify the mark because the mark was not created within the current Studio Session.

So, in its current state, utilizing Sessions with Sets requires a lot of work and doesn't behave as well as one might expect. That said, the authors believe it would be a great step for the future. Having the ability to reap Session functionality on file in a set removes any difficulties associated with typical Windows server limitations and read-only files.

A few examples of specific capabilities are as follows:

- Ability for an administrator to perform content level changes like Digital Slip Sheeting

- Feature enabling a match of all existing markups in a newly uploaded Studio Session document with all the users in the Studio Session, similar to the Digital Slip Sheet match, thus allowing users to modify their markups

- Ability to see files in a similar format to a set within a Studio Session

The second big improvement the authors would like to see is a document status feature. Following the trend and moving in the direction of more document management, adding a file or sheet status could significantly improve workflows.

Figure 11-1 is a conceptual sketch of how the feature might work. Note that each sheet in a set has a status indication in the bottom right corner. Further, the status indication has a dropdown arrow allowing the user to select a status for each sheet. The statuses might be "Reviewed," indicating the sheet has already been reviewed, or "Incorporated," indicating redlines or markups have been picked up and the sheet is ready for backcheck. The sheet statuses would help professionals know which sheets require their attention without actually opening the sheet.

Further, a filter capability would help professionals create a to-do list. The filter would work just like the markups list and only show sheets meeting the filter requirements. So, an engineering technician could filter for all sheets with a "Reviewed" status and quickly determine which drawings he or she need to update.

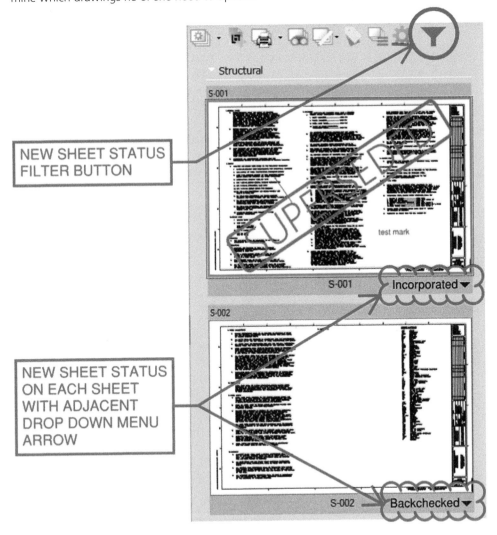

NEW SHEET STATUS FILTER BUTTON

NEW SHEET STATUS ON EACH SHEET WITH ADJACENT DROP DOWN MENU ARROW

Figure 11-1: Conceptual Sheet Status Sketch

Geo Location Features

With the recent boom of Geographic Information Systems, or GIS, it seems that PDF and Bluebeam specifically should start considering what GIS-type data could be captured.

One possibility that seems to have good potential would be the ability to utilize a tablet's GPS feature to geo-locate a point on a PDF. Utilizing a plan view drawing, a user could physically walk to a location and click the corresponding location in a drawing, from which Revu would then scale and locate the plan view in space.

Once located, the user could associate photographs with locations, no longer having to mark or locate photos taken in the correct place; instead, that would happen automatically. So if a field inspector was taking photographs of a damaged foundation to initiate an RFI, the photo would know where to go on the plan view based on the location of the iPad. One step further would orient a camera icon in the direction of the camera lens.

Another possibility would be to use the PDF as a map, with a "current location" icon that would work similarly to the "follow" function. When turned on, the user could see where he or she was standing on the plan view. Stretch that further and a project manager might even be able to know where every professional on the jobsite was located. In hazardous situations, this could be a life-saving development because professionals could be easily accounted for as long as they carry their iPad with them.

Once again, the possibilities seem endless. With no such integration currently developed, the sky truly is the limit.

Virtual and Augmented Reality

The final thought on future capabilities is influenced by virtual and augmented reality (VR/AR). The industry as a whole is quickly realizing that VR/AR is not just for gamers. Professionals are scrambling to figure out what they can do to use this technology in a valuable and meaningful way. Some ideas have to do with live virtual model reviews, while others have to do with noticing errors in construction. But what does all this mean to Revu?

Can the PDF specifications integrate with VR/AR? Maybe they can, or maybe they don't need to. What if Bluebeam embedded their powerful markup tool into a VR/AR platform? Instead of marking on sheets and 3D models, users inside of a virtual reality experience could mark on walls, equipment, and objects. Instead of jumping to a mark on a drawing, clicking a line in the markups list would teleport the user to a new location. And text wouldn't be typed on a customary keyboard; it would be spoken, simplifying the experience for the user.

Take that idea one step further and model review meetings could happen inside a virtual environment. Multiple users could teleport to the location of a comment and then discuss it while standing in the model, looking directly at the issue at hand, all while taking notes in Revu.

Is it realistic? It probably is. Would it be cool? Most definitely.

Conclusions and Final Thoughts

Revu became one of both authors' favorite tools almost immediately and has only improved its ranking in the years since. In addition to being a favorite, it is also one of the most used programs in our arsenal, second only to Microsoft Outlook. As engineers, that's a funny thing to say because as a student going into the profession, I expected to spend most of the day designing, not communicating, but that expectation is dead wrong.

In its relatively short time in existence, Bluebeam has made a splash in the industry that is second to none. The growth in users and transformation in capabilities and features is somewhat remarkable. Even with all that history, there is still the question of whether it has staying power. Will Revu be able to remain relevant in the wake of the ever-quickening pace of technology?

With the growth and improvements of BIM, will redlines and markups cease to be completed on sheets and PDFs? Will construction documents become construction models with radically different review methods and building techniques? Will artificial intelligence, robotics, and additive manufacturing automate the entire process, eliminating the need for a tool like Revu? Will we stop building in the physical world, preferring the experience of virtual reality?

Again, we can't predict the future, which is too bad, but true. What we do know is that Bluebeam is a company built on the foundation of community, both internal and external. It is a community the professionals at Bluebeam work hard to connect and grow. The connection between the company and their customers is unparalleled. They are always listening, always innovating, and always supportive. Did you know that e-mails to suggestions@bluebeam.com go directly to the product management team? Now you do!

Knowing many professionals from Bluebeam, their passion for what they do, and the transformation they've made up to now, we find it hard to imagine a reality where Revu becomes irrelevant. Instead, we see a bright and promising future. Revu may not look or function like it does today—in fact, I'll almost guarantee it—but the same is true of Revu today compared to 10 years ago.

If the team continues that strong customer connection, Bluebeam will adapt to the needs of the times and adjust to the changes technology brings. We can hardly wait to see what's next because it seems that each time it gets more exciting.

We hope this book has given you something to take with you, whether that's one new skill, a bunch of new ideas, or simply the courage to change. When we started this book, we set out to share both what we have learned and the stories of people we have met. The stories are some of our favorite examples of professionals who have defied the odds to do something really exceptional in our eyes, each individual brave in his or her own right.

Of course, we don't want the stories to stop here; we want you to share your story with someone else. As we noted in the introduction, we, the authors, are better thanks to the many people who have shared what they know with us. Together, we can go faster, dream bigger, and aim higher.

Index